Facilities Management and the Business of Space

Facilities Management and the Business of Space

Wes McGregor & Danny Shiem-Shin Then

Wes McGregor is a leading space planning consultant
and a Director of the AWA

Danny Shiem-Shin Then is
Associate Professor in Strategic Asset Management and
Facilities Management at Queensland University of Technology

ARNOLD

A member of the Hodder Headline Group
LONDON

Co-published in North, Central and South America
by John Wiley & Sons Inc.
New York • Toronto

First published in Great Britain in 1999 by
Arnold, a member of the Hodder Headline Group,
338 Euston Road, London NW1 3BH

http://www.arnoldpublishers.com

Co-published in North, Central and South America by John Wiley & Sons, Inc.
605 Third Avenue, New York, NY 10158

British Library Cataloguing in Publication Data
A catalogue record for this book is available from the British Library

Library of Congress Cataloging-in-Publication Data
A catalog record for this book is available from the Library of Congress

ISBN: 0 340 71964 8
ISBN: 0 470 38170 1 (Wiley)

1 2 3 4 5 6 7 8 9 10

Publisher: Eliane Wigzell
Production Editor: Julie Delf
Production Controller: Iain McWilliams
Cover Design: Mouse Mat

Typeset by Phoenix Photosetting, Chatham, Kent
Printed and bound in Great Britain by St Edmundsbury Press, Bury St Edmunds, Suffolk and MPG Books, Bodmin, Cornwall

Contents

Preface

The world is an ever-changing place. In 25 years the world of work will look very different from the way it looks today. And while such a statement could have been made at any time in history, what singles out the dying hours of the 20th century is that the extent of change is much more than it has ever been before – dramatic economic shifts, revolutionary technological development, new visions of how business should be run and how organizations should be shaped to respond to the changing dynamics of business challenges.

Much of the workspace inhabited by people today has at best been developed in response to assessments of work needs that have their origins in a time when the pace and character of changes in work were much less pronounced than they are today. The workplaces of the future must not only accommodate rapid changes – economic, technological and social – but must also strive to reflect and promote new ways of working.

Work environments in terms of their location, size and configuration are, or at least should be, a direct response to the considered needs of people and the work processes they carry out, or plan to carry out in the future. Consequently, to ensure that the most fitting work environments are provided to meet the needs of business, those tasked with the responsibility of developing workplace strategies and planning workspace must be constantly considering their needs for the years ahead.

The facilities manager is unlikely to be involved in helping create the future shape of the business and the organization. 'It is not their job', is the frequent response, 'it is the work of others'. While it may be true that strategic visioning, market positioning, and organizational planning are within the realms of core business managers, it is also undoubtedly true that the quality of the work environment provided has the ability to enhance or detract from the performance of those who work in it. Furthermore, it is the responsibility of those who are tasked with planning the future direction and strategic intent of the business, to consider the means by which the organization should deliver its goods or services to customers, i.e. the tools they need. If the emerging role of facilities managers is to make an impact in adding value to the business delivery process while minimizing cost, then these managers must aspire to a position of influence via their ability to offer facilities solutions to address business challenges. Such a position of influence can only be attained by actions that are in tune with the strategic intentions of the organizations they are serving.

In the old way of planning workspace, the temptation was to overestimate the amount of space in the knowledge that eventually, as the volume of work increased and head count grew with it, the initial excess of workspace would be absorbed. In such a world the facilities manager may even have been considered to be a saviour as he or she had the 'foresight' to take on the extra space. However, today the world is different. In an age of rapidly changing business needs, new ways of 'employing' people and a different geography of the marketplace, facilities managers can no longer rely on the inevitable increase in headcount to absorb their planning mistakes. Add to this the pressures most organizations face for reducing costs, especially the unit cost of production in both manufacturing and service industries, and the need to reduce the reliance on, and impact of, fixed costs such as rent and rates, then it is clear that facilities managers and their businesses need to consider their workspace in ways different from those of the past.

All too frequently the results are well intentioned compromises, made in an effort to address the needs of the business in premises that are no longer, if they ever were, fit for purpose. This is not an indictment of facilities managers' capabilities, but a recognition that their efforts in so many cases were misdirected in attempting to overcome corporate obstacles placed in their path. Often, these obstacles are the result of poor or inadequate consideration of workspace by those charged with planning the direction of the business, i.e. it was regarded as an afterthought. Yet it must also be acknowledged that many facilities managers have no greater desire or wider focus, other than addressing issues relating to the premises in which the organization currently works. So how does this mismatch occur? And what can be done to address it appropriately?

The proactive alignment of real estate asset management and workplace strategies with corporate goals and the objectives of business units, will continue to be the push for facilities managers; their focus being to add value and minimize the operating costs of the business processes they are supporting. In the overall management of the provision of corporate workspace, facilities managers of the future must strive to assume the mantle of 'leader of the workplace'.

This book, born out of the authors' experiences in research and practice, is intended to assist those for whom the planning of an organization's workspace is a key component of their job. It looks at the changing nature of work and the implications for the provision of the workplace – in short, the demand and supply sides of the workspace equation. We trace the growing trends of organizations that are integrating their physical resource considerations (built assets) into their overall business planning processes. We describe methods that can be employed to gain an understanding of the needs of the business (current and future) and the capabilities of current workspace, and how to reconcile the supply and demand equation. We aim to show the correlation between work and the workplace, between workplace provision and the effectiveness of people, and the correlation between the way the workplace is provided and serviced, and business's effectiveness in meeting the needs of their customers. Finally, we also encourage 'stepping back from the canvas' – that is to say, putting the details of your organization's forecast workspace needs into the context of the changing world of work – and in so doing create an opportunity to add real value to the organization by moving from *responsive* to *anticipatory* workspace planning.

In each chapter we use Case Notes, drawn from businesses we have encountered through our research and consultancy work, to illustrate the topics covered and which we hope will exemplify and amplify the text.

This book covers many topics relating to buildings, technology, environmental services and their management. Essentially, however, this book is about people. That is, the people who design, construct, service, maintain, lease, buy and sell workspace – but most importantly of all, it is about the people who use workspace. It is about their needs and aspirations, their likes and dislikes, and how to create workplace environments that best support them in their endeavours.

Acknowledgements

This book could not have been written without input from a great many people made in a host of different ways, without their help and guidance this book would not be what it is today.

We wish to thank all of those who have helped us and we especially wish to acknowledge the contributions of the following people: Professor Peter Barrett, Salford University; Clive Hamond, Regus; Dr John Hink, Heriot-Watt University; Andrew Mawson, Advanced Workplace Associates; Neil McLocklin, BT Workstyle Consultancy; Hugh Anderson, Hugh Anderson Associates; Nick Philips, Lloyd's; and Paull Robathan, Enigma FM.

For their assistance with the photographs that accompany the text, we would like to thank: Ann Inman, Claremont Office Environments 1.2(a), 1.2(b), 2.4(a), 2.4(b), 4.10(a), 4.10(b), 4.2, 6.9; British Telecom Picture Library 4.8, 8.4, 8.6; Martin Elliot, Spring Mobile Training 6.10(a), 6.10(b); Tom Harvey and Mike Clift, Building Research Establishment 5.10, 5.11; Crispin Boyle 6.15; Ros Webb, Talisman Marketing 4.3(b), 5.16(a), 5.16(b); Eddie Brinsmead-Stockham, Microsoft 6.13; Armelle Dhieer, Steelcase Strafor 1.2(a), 1.2(b), 2.4(a), 2.4(b), 4.10(a), 4.10(b); Angelique Mooijweer, Regus 5.5; Dennis Beitz, Beitz Alberti Consulting Group 5.14; Tony Fowler, Nationwide Building Society 6.15; Nick Philips, Lloyd's 2.5; Amanda Aspinall and Tony Leppard, Decision Graphics 3.9; Hugh Anderson, Hugh Anderson Associates 8.8(a), 8.8(b), 8.8(c); and Paul Zanre 1.1(a), 1.1(b), 2.6(a), 2.6(b), 3.6, 4.3(a), 4.6, 5.5, 6.17, 6.21(a), 6.21(b), 7.6(a), 7.6(b), 8.5, in addition to which Paul's guidance and enthusiasm were much appreciated.

We must also acknowledge the inestimable contribution made unknowingly by all those people who, as clients, colleagues and friends have, over the years, helped shape our thinking and our actions leading eventually to this book.

Thanks also go to our publishers for their flair and backing, particularly Eliane Wigzell for her guidance and patience.

For their encouragement and understanding throughout the days and many nights of work, we owe more than just thanks to our respective wives Cath and Fui-lan, without whose unstinting support this book simply would not have happened.

We hope we have acknowledged everyone who has given us assistance with this book, and apologize to any we may have inadvertently overlooked. While many people have contributed to the development of this book, the opinions and statements made, right or wrong, are ours alone. While every effort has been made to establish the authenticity of the information we have obtained, if there are any inaccuracies or omissions in the text, responsibility for such occurrences rests solely with us.

Wes McGregor
Fearnan, Scotland
March 1999

Danny Then
Brisbane, Australia

Glossary of terms

The following terms are used throughout the text of this book, and any reference made to them is in the context of the definitions ascribed to them below.

Ancillary space The usable area of workspace given over to functions that support an individual department of working group, i.e. filing, local machine areas, meetings, departmental storage, shared terminals.

Churn A measurement of change of workspace, where the number of workplace moves carried out in one year, is expressed as a percentage of the number of workplaces provided.

Common support space The usable area given over to functions that support the whole (or substantially the whole) organization or building, i.e. despatch areas, telecom areas, central computer room, library, catering areas.

Contiguous space Uninterrupted floor space available on one floor.

Core area The area containing lifts, stairs, common lobbies, plant and service areas, ducts, toilets and the area of internal structure.

Depth of space The distance between external walls or atrium walls with windows.

Floor plate Gross internal contiguous floor area available on one floor.

Gross Internal Area (GIA) Total floor area of building between internal faces of external walls or atrium walls, including internal structure and core. It excludes roof plant and totally unlit areas.

Net Internal Area (NIA) Total floor area of building between internal faces of external or atrium walls excluding cores (i.e. GIA less cores).

Net usable area The usable area remaining after core and primary circulation have been subtracted from the GIA. It equates to the sum of workspace, common support and ancillary space indicating the usable area on a typical floor plate.

Primary circulation Major routes within the NIA which link stairs, lifts and fire escapes.

Secondary circulation The routes connecting groups of workplaces to the primary circulation route.

Service space The parts of the ancillary workspace which accommodate support service equipment, i.e. photocopiers, printers, fax machines, mail points and vending machines.

Workplace The desk or workstation configuration provided to an individual member of staff, be it enclosed or open plan.

Workplace footprint The area of floor space occupied by the workplace including its attendant access circulation.

Workspace The usable area remaining after support and ancillary areas have been deducted from the NUA. This is the space required for the individual workstations, personal filing and meeting requirements.

All references in the text to the male gender should be regarded as inclusive of the female gender where the context so permits.

Acronyms and abbreviations

The following acronyms and abbreviations, used throughout this book, have the meaning ascribed to them below:

ATM	automated teller machine
BMS	building management system
BRE	Building Research Establishment
BREEAM	Building Research Establishment Environmental Assessment Method
CAD	computer aided design
CAFM	computer aided facilities management
CEO	chief executive officer
CIFM	computer integrated facilities management
CPU	computer processing unit
CRE	corporate real estate
CRT	cathode ray tube
DIP	document image processing
DIY	do it yourself
EDI	electronic data interchange
FPD	flat panel display
FRI	full repairing and insuring
HVAC	heating, ventilating and air-conditioning
HR	human resources
ICT	information and communications technology

IT	information technology
JIT	just in time
LAN	local area network
LBG	Learning Building Group
LCD	liquid crystal display
LEP	light emitting polymers
PC	personal computer
POE	post-occupancy evaluation
PPM	planned preventative maintenance
QA	quality assurance
SFB	strategic facilities brief
SFP	strategic facilities planning
SLA	service level agreement
SLB	service level brief
SoR	statement of requirements
TFM	total facilities management
UMTS	Universal Mobile Telecommunications System
UPS	uninterrupted power supply
VAV	variable air volume
VDU	visual display unit
VR	virtual reality
WAN	wide area network

List of figures

Chapter 1

Chapter 2

Chapter 3

Chapter 4

Chapter 5

Chapter 6

Chapter 7

Chapter 8

List of tables

Chapter 1

Chapter 2

Chapter 5

Chapter 6

Chapter 7

Chapter 8

1

Introduction

'Corporate real estate – the land and buildings owned by companies not primarily in the real estate business – is an aspect of corporate affairs largely ignored in boardrooms across the USA.' So said Sally Zeckhauser and Robert Silverman, writing in the *Harvard Business Review* in 1981. A survey conducted by Zeckhauser and Silverman, president and vice president respectively of Harvard Real Estate Inc, who manage Harvard University's properties, found that only 40% of American companies clearly and consistently evaluated the performance of their real estate, 'most treated property as an overhead cost like stationery and paper clips.'

Almost two decades on, the acceptance by senior management of property assets (real estate) as a vital business resource is still by no means obvious. In many organizations the role of operational property is still considered to be no more than a cost to business, an overhead that does not warrant serious management considerations. This view persisted when the world economy was buoyant and business optimism was bright. The prolonged depression in world trade in the 1970s, and from the mid-1980s onwards, has brought about a renewed awareness in controlling the costs of doing business. For many large corporations, the revelation of the fact that after staff costs (salaries), the next highest category of costs are facilities-related, has reinforced the strategic importance of property (or real estate) as a business resource, and the need to manage the resource as effectively and efficiently as possible.

This new awareness has brought about a much needed management focus on measures to ensure the corporate real estate (CRE) portfolio is matched as closely as possible to operational requirements, and that asset occupancy costs are managed and controlled. The perceived role of real estate assets in business and their effective management is increasingly seen as a strategic dimension in business planning.

The desired outcome in any organization, from the perspective of business operations, is to maintain strategic relevance by attempting continuously to match the demand by business units for workspace, with the existing real estate portfolio (i.e. the supply) through the provision of appropriate enabling work environments that satisfy the needs of the people in the business and their work processes (Fig. 1.1).

The pace of change brought on by intense global competition and rapid technological developments has in recent years meant that the past assumptions of stability and steady growth are no longer true. These changes have particular significance for the ongoing management of operational property, and premises occupancy costs. A key feature within

(a)

(b)

Fig. 1.1 The business park has become a focus for investment in workspace: (a) Edinburgh Park, South Gyle, Edinburgh; (b) John Menzies plc, Wholesale Division Headquarters, Edinburgh Park, South Gyle, Edinburgh.

any organization is the ability to respond to shifts in the strategic direction of the business and, while so doing, control the likely impact upon the existing real estate portfolio, in terms of the amount of space, i.e. the scale of the assets, as well as the financial consequences of its ownership and occupation. Essential prerequisites for this to be achieved are appropriate strategies for both facilities and support services that are continuously aligned with the strategic intentions of the business. This book is an attempt to provide routes to possible solutions, through consideration of conceptual models and the application of practical processes. The aim is to clarify the crucial role played by operational buildings – what we call *workspace* – in supporting the fulfilment of business objectives, which is achieved through the prudent management of their provision and servicing – *the business of space*.

1.1 Buildings and the business of space

During the period up to the mid-1970s, work buildings were seen very much as a necessary, but relatively 'static' factor of production, required to accommodate the production processes of business. Expenditure on these buildings was generally regarded as a 'sunk' business cost that could not be avoided. The prevailing view at that time was that the costs associated with building occupancy were part of the business production process and were therefore necessary business expenses. The level of management associated with controlling this group of facilities-related costs was not regarded as sufficiently demanding nor sophisticated enough to necessitate specialist skills or the attention of the business's senior executives.

The oil crisis of the mid-1970s had the immediate impact of raising awareness of the need to manage the costs of occupancy, particularly those relating to energy. The focus on monitoring energy consumption in occupied buildings had the effect of bringing the 'space' dimension of business premises to the fore. Up till then, the capital costs of building construction had been the major concern in any evaluation of building investments. The energy crisis, therefore, can be said to be largely responsible for creating a climate in which the general focus amongst building owners, business managers and occupiers, was of the need to consider the economics of occupancy in terms of the elements of recurring costs associated with sustaining the operation of business premises, such as energy, repairs and maintenance.

The boom and bust cycle of the property market in the 1970s and mid-1980s caught many companies off-guard. The combination of over-commitments in long-term leases made at a time of optimistic expectations for high growth, followed by a prolonged period of general economic recession and stagnation, resulted in a situation of surplus capacity in real estate provision for many organizations. For many large corporations, the coincidence of a fall in revenue with a consistent rising trend in occupancy costs, resulted in a dramatic shift in the focus of senior management in their efforts to contain business operating costs – getting rid of space surplus to current requirements.

The rapid pace of technological development, particularly in information technology, has, and continues to have, a considerable impact on the design and subsequent use of buildings. Far from being regarded as a necessary evil, as had been thought by many, for an increasing number of organizations there is a growing acceptance that buildings (being operational assets) must now be managed as a valuable business resource, just like people and technology. Further, there is a growing consensus that investment decisions associated with the provision and subsequent management of operational assets, must consider the interplay between:

- **Property** – the physical resource.
- **Technology** – the supporting infrastructure in terms of essential building services, as well as the technology that directly supports work processes.
- **People** – the users of the facilities and services, in the conduct of their value-adding activities.

This acknowledgement of operational property as a business resource has resulted in a growing appreciation of the need to manage the business's operational asset base over time.

The 1980s and 1990s have also seen an enormous growth of service industries. One of the ramifications of this growth, the development of a service culture, has had a direct impact on the property management industry, through the emergence of *facilities management* as a professional discipline, in North America initially, then in the UK and Europe, and most recently in Australia and the Pacific Rim. As part of an emerging service industry, the facilities management market has grown considerably within the property services sector, which by 1991 was estimated to be worth £16 billion per annum (Centre for Facilities Management, UK FM Survey), providing facilities-related support services and management expertise to organizations with large operational real estate portfolios. The growing trend of contracting out (or outsourcing) of non-core support services, far from diminishing the role of management of the organization's real estate asset base, has reinforced the strategic importance of aligning the physical operational asset base to the organization's business plans.

At the same time, there is also a growing acceptance that the workplace environment is becoming a crucial component in the drive to improve productivity of the organization's most expensive resource – its employees. Together with information technology, the management of the operational asset base is increasingly regarded as an important leverage in the strategic management of the workplace environment, aimed at improving overall organizational effectiveness (see Fig. 1.2).

It is within the context of the above developments and their implications for the management of business operations, that the focus of attention is increasingly being aimed at seeking solutions to one key question: how do organizations manage their inherently fixed, operational real estate asset base in a business environment that is increasingly competitive and constantly changing?

In seeking appropriate facilities solutions to business challenges it is important to understand clearly the nature of the product we are dealing with. In terms of buildings as an operational resource, the unit of measure and hence the language of discussion, is the functional space or *workspace*. It is the variables that define the mix of workspace requirements, which must be aligned with the outputs of the business, whether they are tangible products or intangible services. It is in this context that the language of 'the business of space' must be defined, and that the space itself is assessed, procured, serviced and disposed of, as a supporting resource of business.

1.2 Linking real estate and facilities decisions to business strategy

The significance of the strategic role of real estate assets to corporate performance, can be seen in the influential research report by The Industrial Development Research Foundation (Joroff *et al.*, 1993), where corporate real estate assets are termed as *the fifth resource*, after the traditional resources of people, technology, information and capital.

There is clear evidence (Zeckhauser and Silverman, 1981; Veale, 1989; Gale and Case, 1989; Avis *et al.*, 1989; Arthur Andersen, 1993) that supports the view that an assured strategic direction is needed from senior business managers, in their consideration of the

(a)

(b)

Fig. 1.2 (a) and (b) The importance of the work environment on business effectiveness and hence profitability, is gradually being recognized.

management of corporate real estate (i.e. the land and buildings used for workspace, infrastructure and investment) as an integral part of business resource management. This view was aptly expressed in the Corporate Real Estate 2000 Report (Joroff *et al.*, 1993, p.7) which called for a rethinking in the way this resource is managed:

> In the past, corporate management often did not consider the corporate real estate function to be as important as the four corporate resources of capital, people, technology and information. Senior managers had not learned to ask how the function could create value for the company and help to meet the overall corporate mission. Today that goal is being pursued aggressively. Senior managers now are beginning to recognise that real estate is a critical strategic asset, one that supports the financial, work environment and operational needs of the total corporation.

The reference to 'financial, work environment and operational needs' as an integrated task of resource management to provide for the 'total corporation' is significant. It acknowledges that land, buildings and the work environment are essential parts of every corporation's strategic planning, and must therefore be managed accordingly, to ensure that the financial and operational goals of the company are met.

There are many examples where the strategic dimensions of real estate decisions are being demonstrated through the actions taken by real estate or facilities managers in order to add value to the organization, improve economic performance at a time of frequent corporate restructuring, and in so doing, strengthen the company's competitiveness. In most companies, however, the real estate or facilities management function is still primarily reactive and oriented to transactions.

Studies conducted in North America and the United Kingdom in recent years (see Appendix A) all confirmed the relatively low status of real estate and facilities managers, and that real estate assets are an under-managed corporate resource. A key issue confronting real estate and facilities managers in many organizations is how to transform their existing, essentially operational, role into a strategic function. At the heart of the question lies a number of issues that must be addressed:

- The perception the organization's senior management have of the role of the real estate and facilities management functions.
- The reporting line of real estate and facilities managers, and their positioning in the organizational hierarchy of the business.
- Current facilities management processes and their interface with senior management and business units.
- The competencies and capacity of in-house real estate and facilities departments.

The above issues relate to four aspects of organizational variables, namely, *attitudes*, *structure*, *process* and *competencies*. Taken together, these variables will determine the performance of the in-house real estate and facilities personnel as service providers.

In order to link real estate and facilities decisions to the organization's business strategy, the required 'rethinking' mentioned above needs to be applied equally to both strategic management and operational management. It is crucial for senior managers to understand the resource implications of corporate real estate assets in terms of:

- The nature of the physical asset as a enabling facility.
- The economics of the provision of real estate assets as operating resources, as well as in terms of their intrinsic value.
- The relationship between the physical environment, individual satisfaction and organizational productivity.
- The provision and ongoing management of facilities support services and their users' interface with the workplace environment.

Operational management (i.e. provided by the real estate and facilities function), for its part, has to move from being a transactional-reactive role to a strategic-proactive role, where the emphasis is on anticipating the future in light of the company's core business and its work processes, and in consistently providing value-adding solutions. Congruent with senior management's 'new awareness' of corporate real estate assets as described above, real estate and facilities managers will have to be able to provide innovative solutions to:

- Meet competitive challenges, for both the business and for the facilities management function.
- Reduce costs associated with acquiring, operating and changing workspace and its layout.
- Increase the quality of the work environment and hence improve the productivity of its users.
- Respond to unpredictable market conditions and, in so doing, have close alignment of the workspace provision with the needs of the business.
- Organize workplaces to accommodate a more varied and complex workforce.
- Exploit new information technologies, to improve the management of workspace and its related facilities services.

It should be noted that the *cost* and *quality* referred to, are not alternatives. The days of trading one aspect off against the other are gone – what business expects, in fact demands, is low cost *and* high quality. An *informed interface* between strategic management and operational management is clearly needed, in order to address overlapping concerns and to reconcile top-down communication of strategic intent with bottom-up performance reporting of the real estate asset management function. Management development, promoting this informed interface between corporate planners and real estate and facilities managers should be driven by a clear motivation that balances:

- the demand to control costs and minimize long-term commitments to infrastructure – both of which suggest the consumption of less functional space; with
- the increasing need to provide workplaces that enhance productivity, while addressing increasingly complex environmental requirements; and
- the provision of satisfaction to the workforce, individually and collectively.

Management responses to meeting these strategic requirements have evolved as a form of strategic planning – strategic facilities planning – which we will consider in the next chapter. The emergence of strategic facilities plans as corporate development tools, linking real estate decisions to corporate business plans, supports moves toward proactive management of the corporate real estate portfolio. The trend is endorsed by numerous studies over the last decade, which include Varcoe (1991a, 1991b), Debenham Tewson Research (1992), Arthur Andersen (1993), Avis *et al.*, (1995), each of whom stressed the need to consider the real estate resource more strategically in future business planning.

1.3 Space as a business resource

From the foregoing discussion, it is discernible that a fundamental rethinking is required about the definition and the use of space. A key concept in the measurement of the performance of operational real estate assets, is the relationship between the cost of provision and the utilization of workspace. Workspace, measured in terms of square metres or square feet, is a component of the real estate resource. The monitoring of occupancy costs as a primary component of the costs associated with facilities provision, has focused management attention upon the importance of the amount, quality and utilization of workspace, as the key business measures of the real estate resource. Defining workspace as a business resource demands a clear understanding of the organization's operational support needs at two levels.

- Corporate real estate assets at the portfolio level – attributes, location and tenure.
- The characteristics of individual premises at the building level – attributes, floor plate and layouts.

At the portfolio level, matching supply to demand is, or should be, an integral part of the interface between core business planning (i.e. the client) and strategic facilities planning. At the building level, the process comprises essentially space planning and space management issues, involving the interface between business units (i.e. the customers) and the real estate/facilities function (i.e. the service provider). The nature of these tripartite interfaces between corporate management (where executive decisions to allocate corporate resources rests), business units (who are the purchasers of facilities and services), and the real estate/facilities function (who are the enablers), suggests that the management of facilities and services provision must be considered as part of a composite whole, and not in isolation. The imperatives of such an interrelated process are:

- In the provision of operational real estate – an increasing realization that the real estate resource can be managed to promote organizational change; and further, that productivity of staff, the business's most expensive resource, can be enhanced by the provision of an appropriate enabling working environment.
- In the provision of support services – an increasing awareness of the requirement for a more systematic approach to defining service levels, and in the procurement of facilities support services upon which a business can develop and evolve.

From all of the foregoing it is evident that a need exists for a hitherto absent dialogue between business managers and facilities managers, based upon a common language, shared objectives and a free exchange of ideas, from which overall benefits can be derived through the effective provision and servicing of the corporate work environment.

1.4 The structure of this book

The objective of this book is to provide an informed discussion of a facet of facilities management that is gaining significant exposure in many organizations in the USA, Europe, Australia and the Far East – that is the management and planning of space.

The level of awareness of space planning and management issues amongst corporate management in many organizations (especially professional practices, financial institutions and banks) has risen greatly in recent years, largely as a result of the need to reduce costs as an agent of improved competitiveness. For many organizations property-related costs are second only to the cost of labour. The drive for overall efficiency and improved productivity in most businesses, has led to a growing awareness of the need to provide *quality* in space provision, rather than the traditional emphasis on *quantity* with its inherent cost implications. In short, the late 1990s saw emerging signs of a greater willingness than ever before, on the part of businesses, to look at the economics of space provision at the strategic level. In turn, this has resulted in the steady demand for commercial training courses, conferences and seminars on the subject, as well as the inclusion of space planning and workplace strategies as integral parts of most undergraduate and postgraduate courses in facilities management.

This book aims to provide a framework for explaining the 'business of space' from the context of business processes and their sensitivity to changing technology and market dynamics. It places emphasis on evaluating space demand as a first stage in the overall process of procuring a building to support business operations. The effective management of space within buildings is seen as the vehicle for enhancing the working environment to support an organization's most expensive resource – its people.

In this chapter we set the scene by emphasizing the inextricable link between people, buildings and business processes. The business of space is about the crucial role of operational buildings in supporting the fulfilment of business objectives by prudent management of the provision and servicing of workspace.

In Chapter 2, we define space management and make the case for businesses to plan their workspace in terms of the type, amount, attributes, location and quantity to meet their needs through time.

Chapter 3 describes the main components that a space planning process should possess in order that it delivers successfully the objectives of the business.

In Chapter 4, we consider the needs of the organization in terms of the demand for functional space. Here we focus on the types of information required for demand assessment; what data are required, how the data are captured, and how the data analysis can lead to the development of workplace solutions that truly support the business and the people within it. We look at the types of information required, work and work processes, the geography of the workplace, workspace allocation and standards, and the process of forecasting future needs.

The supply side of the space equation is explored in Chapter 5, by assessing the key variables that determine the continuous suitability of the existing building, to meet the business's demand for functional space. The building audit will cover issues relating to location, condition, utilization and facilities support infrastructure.

In Chapter 6, we discuss the nature and implications of managing change. We review how these factors impact on the dynamics of matching demand and supply, in order to achieve a successful outcome in terms of facilities solutions that add value to the business delivery process.

Managing space demand over time is the focus of Chapter 7. Given the turbulent nature of the business environment today, the issue of sustaining strategic relevance in terms of the organization's accommodation strategy is of paramount importance. Defining and achieving flexibility in facilities provision (supply) and effective control of ongoing management of churn (demand) are issues for which facilities managers must develop coping strategies.

Inevitably, the future of workspace management will be influenced, if not determined, by what the future holds for the component parts of the space management process. In the concluding Chapter 8 we therefore explore what the future holds in terms of work and work processes, the people who comprise the workforce, their lifestyles and expectations; the technology of work affecting both the users of the workspace as well as the process of its management; and the property industry and its responses to the dynamic needs of those it serves. In particular, it is significant to postulate, if these changes occur, what impact they could have, individually and collectively, upon the role of facilities managers and the process of managing a business's workspace.

In structuring the flow of this book, we are conscious that we have presented the process of the planning and management of space, as discrete phases in the change process affecting the design, planning, implementation and servicing of workplace solutions. Transforming the corporate workplace to meet new and emerging business challenges is a complex and continuous process. It requires strong leadership and a clear vision, a thorough understanding of business processes and how they are likely to be impacted by technological innovations, culture and lifestyle developments. Managing such a complex workplace transformation is hardly a clearly defined, linear process. In practice, linearity is the exception rather than the rule, where the change process is typified by revolution rather than evolution, with overlapping and iterative activities.

Case Note 1a

British National Oil Corporation (BNOC), Glasgow and Aberdeen. In 1980, just over 4 years after its formation, the state owned BNOC embarked upon the development of a plan for its operations in the UK. The events thereafter are as follows.

- In early 1981 development plans are agreed for the construction of new facilities in Glasgow and Aberdeen. The plans envisaged the construction of a new freehold headquarters in Glasgow, to be developed in phases up to a total of 46,500 sqm (NUA), and a new operations centre in Aberdeen, to be developed on a leasehold basis in phases up to a total of 37,000 sqm (NUA).
- Construction of the first phase of the Aberdeen development (18,000 sqm) commenced in May 1982, and was completed enabling occupation in November 1984.
- In August 1982, BNOC is privatized and its name is changed to Britoil.
- The first phase (30,000 sqm) of the Glasgow headquarters commenced construction in September 1983, was opened in December 1986, and fully occupied by February 1987.
- The second phase (15,000 sqm) of the headquarters commenced construction in January 1987 and, although completed in August 1988, was never occupied.
- Britoil was acquired by BP in 1990, the Glasgow and Aberdeen buildings are retained as operations centres.
- 1992 sees the Glasgow building vacated by BP and sold in 1994 to financial services company Abbey National, for occupation by its life assurance and pensions subsidiary Abbey Life who had just acquired Scottish Mutual.
- In 1994 BP negotiate with its landlords to exit their lease of the Aberdeen premises.

In summary therefore, within 15 years a development plan was prepared for 83,500 sqm of accommodation, two buildings amounting to approximately 63,000 sqm (NUA) and costing over £52 million, were designed, constructed, occupied, vacated, sold or had their lease surrendered. Was it worth it? What did the business gain, and at what real cost? Was there another way that it could all have been handled to better effect for the business?

References

Arthur Andersen (1993) *Real Estate in the Corporation – The Bottom Line from Senior Management* (Arthur Andersen) pp. 7–8.

Avis, M. Gibson, V. and Watts, J. (1989) *Managing Operational Property Assets* (Reading: Graduates to Industry).

Avis, M. Gibson, V. and Watts, J. (1995) *Real Estate Resource Management* (Wallingford, Oxon: GTI).

Debenham Tewson Research (1992) *The Role of Property – Managing Cost and Releasing Value.* Debenham Tewson.

Gale, J. and Case, F. (1989) A study of corporate real estate resource management. *Journal of Real Estate Research.* **4** (3), 23–34.

Joroff, M., Louargand, M., Lambert, S. and Becker, F. (1993) *Strategic Management of the Fifth Resource: Corporate Real Etate. Report of Phase One: Corporate Real Estate 2000* (USA: The Industrial Development Research Foundation).

Varcoe, B. (1991a) Proactive premises management – the premises policy. *Property Management.* **9** (3), 224–230.

Varcoe, B. (1991b) Proactive premises management – asset management. *Property Management.* **10** (1), 224–230.

Veale, P. R. (1989) Managing corporate real estate assets – current executive attitudes and prospects for an emergent management discipline. *Journal of Real Estate Research.* **4** (3), 16.

Zeckhauser, S. and Silverman, R. (1981) *Corporate Real Estate Asset Management in the United States* (Cambridge Massachusetts: Harvard Real Estate).

Appendix A

Major survey reports on management of operational property – UK and Europe

Source and title of report	Survey sample	Main survey findings
UK and Europe Surveys		
Managing Operational Property Assets Avis *et al.* (1989).	230 organizations (28.75% response rate, evenly split between private and public sector)	The overall picture was one of reactive then proactive property management. There was clear evidence that property was only seriously considered by organizations when they were under severe profit or cost constraints.
The Role of Property – Managing Cost and Releasing Value Debenham Tewson Research (1992)	Based on interviews 100 major companies.	Only on rare occasions does property receive explicit treatment in corporate plans. More often than not property is viewed as incidental, as an asset that requires little management, generates cost but has little or no value.

Appendix A (continued)

Source and title of report	Survey sample	Main survey findings
Property Management Performance Monitoring Oxford Brookes University & University of Reading (1993)	In-depth case study of the property monitoring procedures of three large organizations	The report cautions that the whole area of monitoring organizational property assets is relatively new and recent research efforts are necessarily a first step towards understanding best practice and therefore, are not definitive.
The Property Cycle – The Management Issue Ernst & Young (1993)	Combination of telephone and interviews. Senior executives of 61 organizations comprising of three types of organizations: property funders, developers and a range of occupiers of commercial property	Though nearly half of the occupier respondents considered the cyclical progression a major shortcoming of the property market, few had a property strategy for their operations that amounted to more than 'we'll find the space when we need it'.
Property in the Board Room – A New Perspective Graham Bannock & Partners Ltd, Commissioned by Hillier Parker (1994)	Personal interviews with 12 finance directors and other directors of UK companies from the private sector. A postal survey of 111 property managers.	Apart from companies whose core businesses are directly tied to the property assets, the perception of the role of property is seen as a cost of business rather than a business resource that requires strategic attention by senior management.
Real Estate Resource Management Avis *et al.* (1995) GTI.	155 organizations (>25% response rate, evenly split between private and public sector)	The findings establish that the real estate resource for organizations in the study is extensive, complex and dynamic. However, real estate still lags behind other key resource areas in terms of attention given and performance achieved.
Shaping the workplace for profit Gallup. Commissioned by Workplace Management (1996)	200 Financial Directors and Managing Directors/ CEOs across a broad spectrum of businesses.	Given the general acknowledged link between the workplace environment, employee satisfaction and profitability, senior managers do appear to be missing an opportunity to manage the working environment for competitive advantage.

Appendix A (continued)

Source and title of report	Survey sample	Main survey findings
The Milliken Report: Space Futures The Henley Centre. Commissioned by Milliken Carpet (1996)	Telephone survey of 200 facilities managers and 50 architects/designers, plus 10 in-depth interviews.	Space management must play a bigger part in overall business development, becoming a strategic rather than an operational issue.
Wasted Assets? A Survey of Corporate Real Estate in Europe Arthur Andersen (1995)	20 companies in eight European countries in three sectors: financial services, manufacturing and retail/distribution. Based on interviews with property executives.	Many companies are missing opportunities to reduce cost and enhance performance because they give limited attention to managing their property assets.

Main survey reports on management aspects of operational property – North America

Source and title of report	Survey sample	Main survey findings
North American Surveys		
Corporate Real Estate Management in the United States Zeckhauser and Silverman, Harvard Real Estates Inc (1981)	300 US companies, 22% response rate. Multi-sector survey.	Despite their enormous value, corporate real estate assets are under-managed.
Managing Corporate Real Estate Assets: A Survey of US Real Estate Executives Veale (1987) Massachusetts Institute of Technology	284 organizations not primarily in the real estate business, 15% response rate.	The under-management of corporate real estate assets is hampered by lack of adequate information on the real estate portfolio and senior management regarding real estate as a cost rather than as a resource.
Real Estate in the Corporation: The Bottom Line from Senior Management Arthur Andersen (1993)	726 US and Canadian companies, 6.2% response rate; plus 50 interviews. Multi-sector survey	Clear differences in the way senior management and corporate real estate (CRE) executives regard the CRE function.

Appendix A (continued)

Source and title of report	Survey sample	Main survey findings
Strategic Management of the Fifth Resource: Corporate Real Estate. Report of Phase One CRE 2000 – Executive Summary Joroff, M. *et al.* (1993) The Industrial Development Research Council (IDRF).	The CRE 2000 research project was partly funded by The Industrial Development Research Council (IDRC) and industrial sponsorships for major Fortune 500 corporations in North America. Input into the project was provided by a broad-based group known as the CRE 2000 Commission whose membership is select, yet diverse, and represents major corporate real estate leaders as well as service providers and public sector leaders.	This first CRE 2000 report examines the emerging role of the real estate resource in large corporations by: • interpreting the impact of corporate change on the requirements for real estate strategies; and • identifying innovative strategies that corporate real estate leaders are using to organize in order to align the real estate resources to fulfil corporate objectives. The report produced a five-stage corporate real estate developmental descriptive model.
Corporate Real Estate 2000. Phase Two CRE 2000 Reports: 1. Reinventing the Workplace Joroff and Becker 2. Toolkit: Reinventing the Workplace Joroff *et al.* 3. Generating High Performance – Corporate Real Estate Service Lambert *et al.* 4. Decision Support Cameron and Duckworth 5. Managing the Reinvented Workplace Sims and Joroff The Industrial Development Research Council (IDRF).	As above	Phase Two of the CRE 2000 Research produced five separate reports which focus on disseminating 'best practice' tools. The findings are based on a series of case studies and workshops about state-of-the-art developments in the areas of corporate real estate management.

2

The context of space planning

Our workplace today is designed and managed on the basis of a set of assumptions that have, throughout the 1990s, been the subject of radical change. Increasingly, it is the facilities manager who is often made responsible for the smooth transition from the past to the future. In this chapter, the context of space planning and space management is established within the broader roles of facilities management and strategic facilities planning.

2.1 Space planning in facilities management

Before considering the management of facilities, it is necessary to determine what we mean by *facilities*. We define facilities as '*The infrastructure that supports the people in the organization in their endeavours to achieve business goals*.' In other words, facilities are the *tools* which people in the business have at their disposal to carry out their tasks.

The term 'facilities (or facility) management' has gained widespread use in North America and the UK, has been progressively adopted in Europe and, most recently, in Australia, New Zealand and the Far East. But there is as yet no consistent definition of its scope of activities. The United States of America Library of Congress provides an initial definition that is often quoted to explain the breadth of the field of facilities management:

> The practice of coordinating the physical workplace with the people and work of the organization; integrates the principles of business administration, architecture and the behavioural and engineering sciences.

The definition of the International Facility Management Association (IFMA), clearly implies that a major part of facilities management activities is inextricably tied to the provision and ongoing management of an organization's real estate assets and its facilities-related support services, as a productive working environment:

> Facility management is a distinct management function and, as such, involves a well defined and consistent set of responsibilities. Simply stated, it is management of a vital asset – the organization's facilities. ... Facility management combines proven management practices with current technical knowledge to provide humane and

> effective work environments. It is the business practice of planning, providing, and managing productive work environments.
>
> (*International Facility Management Association*)

The British Institute of Facilities Management (BIFM) adopts a definition that emphasizes the multidisciplinary nature of the role of facilities managers, which includes extensive responsibilities for providing, maintaining and developing services ranging from property strategy, space management and communication infrastructure, to building maintenance, administration and contract management.

> Facilities Management is the integration of multi-disciplinary activities within the built environment and the management of their impact upon people and the workplace. Effective Facilities Management is vital to the success of an organization by contributing to the delivery of its strategic and operational objectives.
>
> (*British Institute of Facilities Management*)

The Facility Management Association of Australia (FMAA) emphasizes the importance of an integrative view of people, processes and the physical infrastructure in order to enhance corporate performance.

> Facilities Management is the practice of integrating the management of people and the business process of an organization with the physical infrastructure to enhance corporate performance.
>
> (*Facility Management Association of Australia*)

The potential scope of activities that are subsumed within the definitions offered by IFMA, BIFM and FMAA can at best be said to be varied. As we will show, *space planning*, that is the organization of the work environment, can be said to be the glue that binds

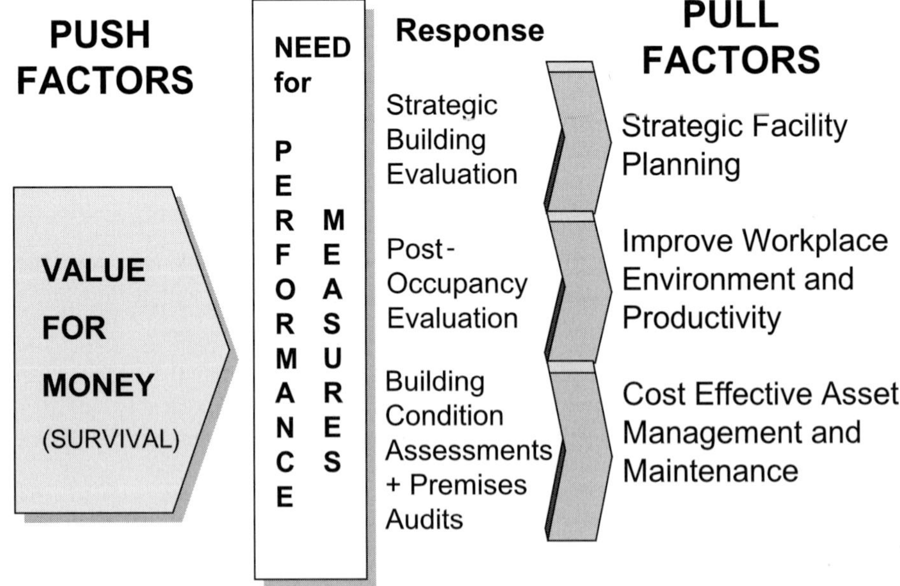

Fig. 2.1 Facets of the facilities management market.

together the diverse and disparate elements of facilities. However, these definitions are not necessarily reflected in the range of services offered by suppliers to the facilities management market, who tend to define their offerings within the narrow confines of the particular services in which they specialize. In general, the market for facilities management services appears to have developed to cater for three interrelated areas associated with the management of operational property (real estate) as illustrated in Fig. 2.1 (Then, 1994b, p.256).

The search for value for money has had one important implication for the whole spectrum of businesses selling products and services of all kinds – the need to describe (specify), to measure (assess performance) and to quantify (price) the output (end product). In terms of operational real estate management, the response for more effective utilization of built assets has been in three main areas:

(1) Strategic evaluation of the real estate portfolio, which has led to the development of *strategic facilities planning* in many large organizations.
(2) Space management and post-occupancy evaluations, which have been driven by the need to maximize utilization of the workplace.
(3) Premises audits and condition assessments, which have raised the awareness and need for cost effective, long-term asset management.

These three developments have prompted a number of writers (for example, Balch, 1994, p.17; Then 1994d,e; Varcoe 1996) to stress the need for corporate management to consider their operational real estate assets as business resources, and to integrate their consideration within the strategic planning process of the business.

In their attempts to define the potential skills base of facilities management, Then and Fari (1992) provided a matrix for classifying tasks that are associated with the property-related aspects of facilities management, as shown in Table 2.1, in which the vertical divisions reflect increasing strategic involvement as we move from a *project tasks* role to an *executive responsibility* role, and where the horizontal divisions reflect the management levels of strategic, tactical and operational.

The range of tasks covered within the matrix may be carried out in an organization

Table 2.1 Matrix of facilities management tasks

	Executive Responsibilities	Management Roles	Project Tasks
Strategic	• Mission Statement • Business Plan	• Investment Appraisal • Real Estate Decisions • Premises Strategy • Facilities Master Plan • Workplace and IT Strategies	• Strategic Studies • Estate Utilization • Corporate Standards • FM Operational Structure • Corporate Brief
Tactical	• Corporate Structure • Procurement Policy	• Setting Standards • Planning Change • Resource Management • Budget Management • Database Control	• Guideline Documents • Project Programme • FM Job Description • Prototypical Budgets • Database Structure
Operational	• Service Delivery • Quality Control	• Managing Shared Space • Building Operations • Implementation • Audits • Emergencies	• Maintenance Procurement • Refurbishment/ Fit-outs • Inventories • Post-Occupancy Audits • Furniture Procurement

either by a facilities manager or by any individual (or individuals) that may not be recognized as being facilities-related. Every item within each cell of the matrix represents a category of decisions that have to be made at various management levels, each requiring appropriate skills necessary to make and implement them, and to assess their effectiveness and performance.

Two aspects that have seen rapid development through the 1990s are:

(1) Asset management and maintenance.
(2) Space planning and management.

Asset management is a term that attracts two distinct meanings depending on whether the term is used in the broad context of investments made by banks, financial institutions and property companies, or in the narrower context of operational property (real estate) management. In the former, investment in property is made for direct profit, while in the latter, the main concerns are with protecting and possibly enhancing the value of the buildings in use. Of course, both approaches are not mutually exclusive. However, our focus here is in the context of the use of buildings, hence the term *operational real estate asset* to denote business use, as distinct from *investment real estate asset*. The importance of taking a whole-life view in the management of maintenance, and components renewal of built assets, is frequently emphasized (Spedding, 1987; Holmes, 1988; Flanagan *et al.*, 1989; Bon, 1989; Marshall and Ruegg, 1990; Then, 1994a, 1995). Each instance points to effective management being derived from the monitoring of the performance of the workspace and the analysis of feedback. So it is that the role of space planning in facilities management is based as much on the measurement of the ongoing effectiveness of the work environment, as it is on the provision of space based upon the assessment of future business needs.

2.2 The relationship of space planning and business planning

Space planning is the professional discipline that incorporates the planning and management of workspace features in many business operations, as diverse as product manufacturing, process engineering, laboratory analysis, retailing and warehousing. In fact, in almost every type of work environment the evidence of space planning can be seen. It is, however, in the business of administration, data processing and related work, i.e. office work, that space planning and space management are under-going their most challenging development. The impact of technological advances in computing and communications, which have recently spawned a number of alternatives to traditional workplace settings, is creating both opportunities and problems for those charged with the management of workspace. The drivers for change in the workplace have not arisen exclusively as a consequence of cost reduction, but instead have been developed from a growing realization that providing a workplace environment for employees, which supports them in their tasks, can lead to improved productivity. This should be the principal corporate objective of providing workspace, a theme advocated forcefully by the likes of Frank Duffy (Duffy and Tamis, 1993; Duffy *et al.*, 1993), Franklin Becker (Becker and Steele, 1995) and Joanna Eley (Eley and Marmot, 1995) among others.

Every journey begins with a single step. The first step in the effective management of workspace is based upon two imperatives – first, the acceptance of the economic reality that every square metre of occupied space has to be paid for by the business; and secondly, that the space occupied by staff should be directly related to the requirements of the work processes and tasks the person has to perform, rather than based on their seniority and status. From this starting point, the facilities manager can embark upon the process of addressing the workspace needs of the business in the full realization that space planning is much more than simply interior decoration and the supply of furniture. It is the conscious planning and design of the work environment to facilitate the delivery of products and services, driven by desired business practices (culture and organization) and operational requirements (tasks and functions). The successful planning and management of workspace is the result of an integrated approach to resource management, which combines a strategic appraisal of facilities provision (real estate assets) with the provision of enabling workplace solutions based upon a rational business case. The justification of the business case for facilities expenditure, is itself based on close scrutiny of premises occupancy costs by business units, as purchasers of functional space. Performance justification is measured in terms of user satisfaction based on a strong customer service orientation within the facilities management team.

'The business of space' is therefore driven by the philosophy that the buildings that accommodate the productive resources of employees, facilities infrastructure and enabling technologies, must be aligned to the strategic intent of the organization they are intended to support. The definition of need, the specification, the procurement and management of functional space as a supporting resource of business, only entered the strategic management arena in comparatively recent times. As the acceptance of facilities management as a professional discipline increases, it will reinforce the crucial role of facilities managers in delivering innovative workplace solutions to meet business challenges of the new millennium. However, this can only be achieved once facilities managers have come to realize that all that is done in the name of facilities management, has but one objective – business effectiveness.

2.3 Business management and economic drivers

Businesses thrive on success which, when measured by purely economic criteria, focuses on efficiency of use of business resources in the production of goods and/or services. This efficiency-driven model of management success owes its origins to the principles of scientific management of Taylor and Gilbreth in the early part of the 20th century. Over the subsequent years, managers have been exposed, it seems almost continuously, to numerous management concepts espoused by the likes of Drucker, Porter, Mintzberg, Deming, Crosby, Peters, Kanter and Handy, to name but a few. In particular, the period since the 1960s has seen management thinking in a state of almost perpetual flux, set against a world economy that has seen wide fluctuations between growth and depression, varying in intensities within and across continents. Writing in *Management Today* Stuart Crainer captures four decades of management thinking as being conceptually characterized as follows (Crainer, 1996).

- 1960s – Strategy is discovered by the corporate monoliths.
- 1970s – The West had nothing to learn from the Japanese.
- 1980s – The radical and rational.
- 1990s – Re-engineering the idea.

The appeal of the concept of 're-engineering the corporation' (Champy and Hammer, 1993) is perhaps, that it best reflects the pace of change over these three decades, and the relentless quest for new ways of managing businesses in order to provide flexibility in an increasingly competitive global marketplace. The authors advocate that businesses must undertake nothing less than a radical reinvention of how they do their work. 'Three forces, separately and in combination, are driving today's companies' they say, 'Customers, Competition, and Change (the three Cs). Their names are hardly new, but the characteristics are remarkably different from what they were in the past.' These three Cs have created a new world for business, and it is becoming apparent that organizations designed to operate in one environment cannot be assumed to work well in another. One of Champy and Hammer's key messages is that change has become constant and is normality, and that the pace of change is accelerating. Acknowledgement of the fact that change has become both pervasive and persistent is of crucial importance in understanding the psyche that prevails in many businesses. This acceptance of the inevitability of change will have direct implications on how organizations adapt to survive and prosper in new, more competitive environments. More importantly, at the organizational level, there are marked implications for how business resources should be organized, procured and managed, to deliver the desired outcomes. Judicious management of all resources – including real estate assets – will continue to be a major factor for companies seeking to maintain a competitive advantage. In the context of facilities management, we are concerned with how change impacts on two distinct but interrelated processes – the provision of facilities and their management over time. The importance of the physical work environment to all modern organizations makes the management of these two interrelated processes an increasing concern for senior managers as they strive to achieve flexibility in their business, cost effectively.

There is an oft-quoted dictum that says – 'business requirements should drive real estate strategies'. However, given the inherent fixity of real estate, companies often find themselves saddled with facilities that once satisfied business requirements but are now inappropriate or obsolete for changing operations. This has been applicable particularly during periods that have seen much restructuring and realigning of business practices. For many international corporations with a large and dispersed operational real estate portfolio, one of the more obvious implications of business volatility is in a downturn following a period of growth. Such occurrences frequently result in the legacy of surplus space from a bygone era, such as prestigious headquarters complexes and long leases in city centre locations, which once were perceived as a necessary, but often hidden, company overhead. Equally, changing corporate requirements for space will continue to generate a compelling need to deal with surplus real estate assets effectively. General Motors and IBM are examples of large international corporations, where the resolution of real estate problems, especially redundant properties that are difficult to sell or recycle, are critical components of corporate survival.

Rapid technological development, particularly in computing and telecommunications, has also rendered many buildings prematurely obsolete, or in need of high capital investment for modernization. In addition, employees' rising expectations of their work environment, coupled with costly and lengthy commuting time to work, are factors that businesses can no longer ignore. In essence, there is a need for organizations continuously to reappraise their real estate and workplace strategies to ensure their alignment to changing business drivers and operational requirements.

In a review of the potential impact of current management thinking on organization structures and the design of the workplace, Duffy and Tanis (1993, p. 427) stated, 'Never has organization theory been so rich and inventive as in the 1990s. Never has the contradiction between managerial aspirations and physical reality been so sharp'. They succinctly placed in context some of the main emerging management themes from the current group of 'new management gurus' (Byrne, 1992) by mapping their implications on the contents of work and the patterns of space use within office buildings, as can be seen in Table 2.2.

This vision of the new workplace, set against a period of accelerating change and intensifying global competition, demands a strategic rethink at corporate level, about the way work is done and the workplace is provided. In turn, this suggests that investments in functional space and technology must be integrated into the strategic modelling of the business, its processes and its varying demands for services and facilities. For example, AT&T's pioneering of the concept of the 'virtual office' for its sales staff, represents an impressive demonstration that real estate and facilities restructuring must work in concert, with both training in cultural changes and systems development support. The concept of the 'virtual office' adopts the principle that, for certain tasks, such as sales staff, the traditional allocation of a dedicated enclosed office to each member of staff does not represent economic use of space. Instead, sales staff supported by portable

Table 2.2 Organizational change and its impact on the physical workplace (Duffy and Tanis, 1993)

The Impact of New Organization Structures on the Workplace	**New ways of working** • more interaction	• more collaboration	• more individual autonomy	**New patterns of space use** • more group spaces	• more shared spaces	• more intermittent space use
Michael Hammer: 'Re-engineering'	•	•		•	•	
George Stalk: 'Time-based Competition'					•	•
David Nalder: 'New Organizational Architecture'	•	•		•		
Peter Senge: 'The Learning Organization'	•	•	•		•	
Charles Handy: 'Discontinuous Change'			•			•
Edward Lawler: 'High-Performance Involvement'	•	•		•	•	
Prahalad & Hamel: 'Core Competencies'	•	•		•	•	
Gerald Ross: 'New Molecular Organization'		•		•	•	
Shoshana Zuboff: 'Informating'	•	•	•	•	•	

equipment (laptops, modems, faxes and phones) can schedule through a reservation system, temporary workplaces and support services, when and where they need them, using a network of offices for contacting customers, administration, report production and team meetings.

In recent years, there is evidence that indicates that some organizations with large operational real estate holdings are beginning to exploit real estate assets as strategic opportunities (Parshall and Kelly, 1993). A well-known example is that of IBM, as described by Sager (1995):

> Rethinking real estate has become an IBM axiom. After shedding 183,000 employees – 27,000 of them in the past year alone [1995] from a high of 406,000 in 1986, IBM simply doesn't need the rambling space it once did. Consolidating space and subleasing, . . . accounted for nearly half of the $6.3 billion in cumulative cost cuts in the past two years. Managers now are evaluated, in part on their ability to shrink property bills.

It is clear that there is a growing realization, albeit mostly among the larger corporations with significant operational property holdings, that their operational real estate can support many facets of business – organizational change, productivity, customer services and communications – if they are aligned appropriately with their business plans. Conversely, if there is a mismatch, where supply exceeds demand, there can be a heavy cost burden and a physical barrier to change. For many organizations, defining their operational demand in physical and spatial terms, is not necessarily an obvious process. This is partly because the process of *facilities provision* is often not seen as a part of strategic business planning, and hence is not operated as such. Very often the decision making processes relating to the provision of operational facilities tend to gravitate towards site selection (location issues) and the building form (design and construction issues), rather than on how the functional space should be designed to support the core business processes (the nature of business). While some organizations explicitly consider how a specific real estate transaction relates to their real estate strategy, this is a rare occurrence. The vast majority of businesses not only fail to make this consideration, they do not even have a formal real estate strategy. A primary reason for this omission appears to be that the real estate strategies that organizations might pursue have not been explicitly articulated or documented. Consequently, most businesses proceed implicitly concerning their real estate requirements, with the inevitable consequences of miscalculation of the amount of workspace required, and/or failure to consider the appropriateness of a specific real estate transaction in the context of the organization's true long-term real estate needs. More seriously, lacking consensus about what the corporate real estate strategy should be, such businesses are likely to pursue transactions without a strategic business context, and consequently risk taking real estate actions that are inappropriate to the strategic requirements of the business.

Surveys of corporate managers (refer to Appendix A) have revealed a curious ignorance and historic lack of interest in relating their operational property (real estate) assets to the overall strategies guiding their business. However, there is some evidence that points to an increasing awareness amongst businesses to consider 'fitness for purpose' at a strategic level. The emergence of this feature has been forced upon many organizations as a result of global competition. One consequence of this is the need to align continuously the organization's physical asset base to its business strategy. In recent years, this has given rise to the development of what is commonly termed as *strategic facilities planning* –

where the emphasis is on aligning the corporate real estate asset base to the strategic intent of the organization, by a systematic evaluation of the following three factors.

- The anticipated demand for workspace.
- The attributes of business critical facilities.
- The operational requirements of each element of the business.

2.3.1 Strategic facilities planning

There have been a number of models employed to illustrate the links between the requirements for physical infrastructure and the corporate strategic plan. The instruments that are used to establish these links have generally been called *strategic facilities plans*. The process of deriving the plans is usually referred to as strategic facilities planning (SFP).

The development of the SFP process as a management tool, can be traced to businesses' responses to changes in global markets and local economic conditions. A period of 'planned growth' during the 1980s was replaced by volatility at the start of the 1990s where change was evident in almost every aspect of business. Pressures to reduce operating costs and increase efficiency, have often meant that many real estate and facilities executives are challenged to provide instant solutions to accommodate any business change. As a result, SFP has developed to become an important modelling tool to input variables and to evaluate various possible scenarios, in order to arrive at the best course of action to pursue. In Fig. 2.2 we illustrate the main input requirements to, and outputs from, the SFP process.

The inputs to the SFP process are clearly multifaceted requiring data from business activities, facilities activities and the financial evaluations of both. Developments in automation and software applications in the field of computer aided facilities management

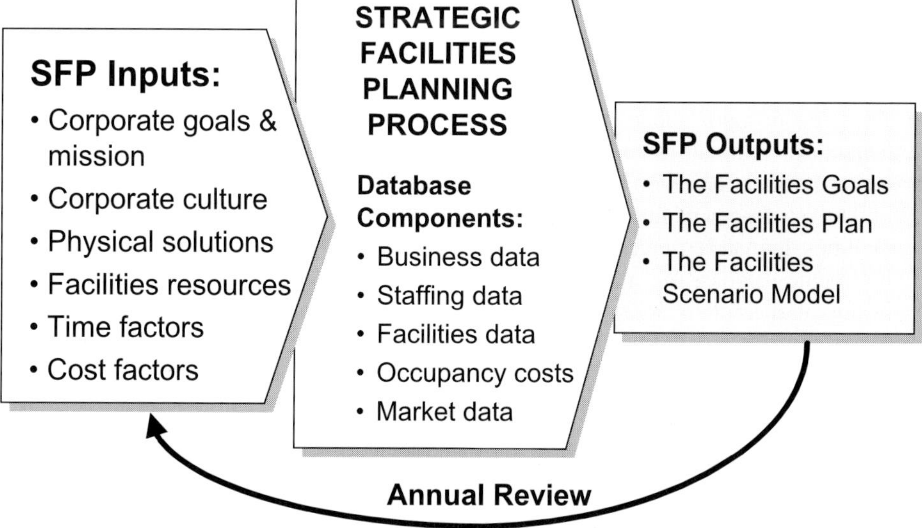

Fig. 2.2 The strategic facilities planning process (adapted from Kimmel, 1993; Brandt, 1993).

(CAFM) and information systems have, in recent years, extended considerably the potential for decision support in space planning and management, especially in scenario modelling. The outputs from the SFP process can be divided into three parts – facilities goals, facilities plans and facilities scenario models.

Facilities goals – show how facilities can help satisfy the strategic business plan. They become the link between the real estate or facilities role and the rest of the organization, including the external provision and management of services. Examples of facilities goals include:

- Enhancement of the organization's public image.
- Development of more productive work environments.
- Attraction of higher calibre employees.
- Improvement of the ability to accommodate workplace changes.

Facilities plans – address what the facilities manager thinks will happen to the company in the next few years. The facilities plans can be seen in terms of evaluating and managing 'current facilities conditions' with a view to meeting the requirements of likely 'future facilities conditions'. In order to fulfil this task, Kimmel (1993) suggests that the facilities plan must comprise three elements – facilities strategies, facilities policies and an up-to-date facilities database.

Facilities strategies – are a set of management decisions that govern the projected use of the business's physical resources for the acquisition, use and disposal of workspace over time. The agreed strategies indicate how the real estate or facilities role will satisfy the established goals, accommodate projected growth and change, and satisfy current needs cost-effectively.

- The implementation of facilities strategies is reflected in *facilities policies* and projects. Examples of policy issues include: own versus lease, build new versus acquire existing, consolidate versus decentralize, in-house provision versus outsourcing, implementing space standards or internal charging. Examples of projects arising from facilities policies could include the revision of space standards, improving space utilization, contracting out design and maintenance services, and the negotiation of revised lease terms.

- The creation and maintenance of a *facilities database* and the use of information technology tools are fundamental to the SFP process, and its capabilities to provide management information and decision support. There is usually a need for integration of data sets from a number of resource perspectives, which could include real estate and facilities, human resources, IT, corporate finance and budgeting. The availability of an up-to-date database and the capabilities of systems are particularly important for the testing of alternative facilities scenarios. As no facilities plan can be regarded as 'fixed' in a dynamic business environment, there is a need for an annual review of the database in order to maintain congruence with the plans of the business.

Facilities scenario models – address what the facilities manager considers could happen to the business. This relatively new component of SFP has evolved largely as a result of the pace of change with which organizations have to cope, and is made possible by the rapid development in information technology, particularly in CAFM systems and, more recently, in CIFM systems. With such tools, multiple scenarios can be established based on possible situations for scenario variables, which may include availability of personnel, budget, outsourcing, business mergers, acquisitions, etc. The scenario model provides the vehicle for multiple evaluations such as the ability of facilities to: support various alternative corporate developments; accommodate different organizational structures and numbers of personnel; and meet varying operational needs over both short and long terms.

The use of SFP has been developed as a catalyst in promoting the much-needed dialogue between corporate planners and facilities managers, as part of the process of aligning real estate decisions to the prevailing business strategy. The guiding principle is that business plans and real estate/facilities plans should be considered simultaneously. SFP represents a new way of thinking about the built environment and its relationship to organizational performance. It is a response to the recognition of businesses that facilities-related costs represent most companies' second largest operating cost, after staff costs, and their largest capital asset. It also reflects that as companies look for opportunities to improve financial performance and competitiveness, a more focused eye will be cast upon the effectiveness of the management of real estate assets and the workplaces they contain, and their ability to support and add value to core business activities. Gradually more and more companies are realizing that their business performance, which is dependent upon the actions of their people, is closely related to how they manage their facilities and workplace assets.

The acceptance of the need to plan strategically for facilities provision, is crucial in developing a shared understanding and dialogue between the two key stakeholders involved in the strategic facilities planning process – business units managers who initiate the demand cycle, and facilities managers who are charged with the delivery of facilities and services, and the ongoing management of workspace.

2.3.2 Functional space as a business resource

The primary driver of demand management of the real estate resource, is meeting business objectives through economic space utilization. The practice of demand management requires a composite approach to resource management, in which the various management disciplines within the business are employed. This view (Then 1994b), regards functional space as the outcome of the synergy of three basic organizational resources – people, property and technology. Figure 2.3 illustrates the scope of management of operational facilities as comprising two main groups of services.

- Facilities-related support services, aimed at providing users with a fully serviced work environment.
- Real-estate-related services, aimed at maintaining the integrity of the building fabric and services (asset management).

The principal role of the facilities manager is to balance the provision and use of the business's interdependent resources of workspace, people and technology with the varying needs of the business. In this respect, the profile of the facilities manager is changing to one that requires more experience in strategic and operational planning, information technologies, finance, accounting, labour relations and communications. One implication of this increased profile is the need to adopt a more analytical and collaborative approach – metrics, models and methods – as well as a creative mindset, in order to evaluate the real

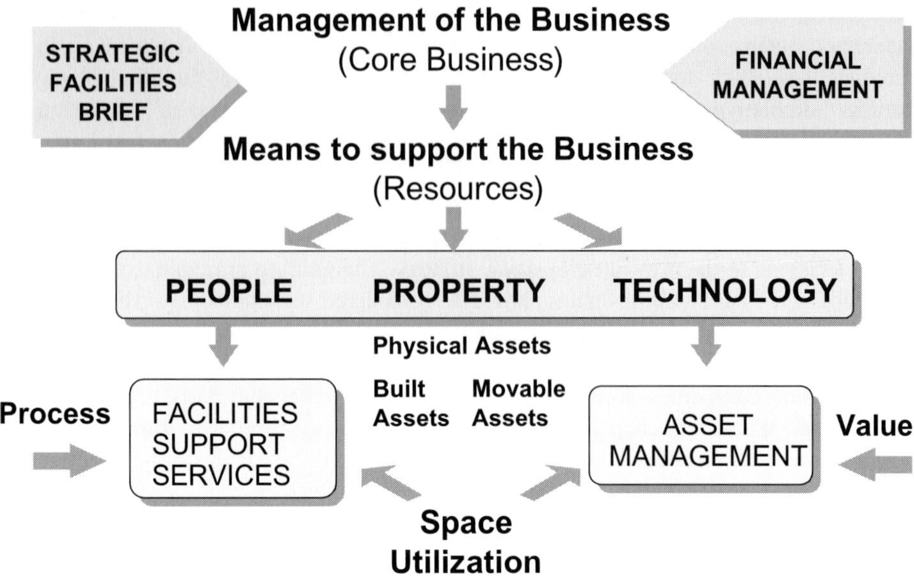

Fig. 2.3 Organizational resources (Then, 1994).

estate portfolio, analyse occupancy costs, measure the impact of property on the business, and formulate a strategic plan. Such tools have a common focus. They attempt to fill the often present void, by measuring the performance of operational facilities, not only in terms of their costs to the business, but more importantly, in terms of their 'fit' to business plans. In so doing, facilities managers can demonstrate how the real estate infrastructure can be managed to facilitate company-wide transformation against a backdrop where change is regarded as the norm.

2.4 The development of the workplace

The selection, development and use of workspace are driven by the collective need of the social unit we call a business, to find an environment in which it can operate. The planning of workspace is a direct response to man's relationship with work and his involvement in work processes. It responds to and shapes buildings and their interiors, it responds to and shapes organizational culture and, most significantly of all, it responds to and shapes people's behaviour. In short, it is both a reflection of society's views and the values it attaches to work, while at the same time acting as an instrument in shaping society's perceptions of work, and how and where it should be conducted.

Analysis of the development of industrial organizations indicates the importance technological innovation has played in the way businesses are structured and hence the workspace they required. So it was that small-scale craft-based production gave way to large-scale mass-production – the workshop giving way to the factory – as first Taylorism (the application of scientific methods to the organization of production), and then Fordism (mass production achieved through standardization and assembly lines to reap maximum economies of scale) became widely adopted.

The type of building we call an 'office', developed from early Victorian times, initially from spaces within buildings, originally designed for use as factories, mills and warehouses. It is only in the late Victorian era that buildings dedicated to office work were created. In these buildings, the work processes involved small groups of people, and in many instance individuals only, writing, copying text and manually producing records of materials, processes, goods, transactions and customers. Clerks sat at large, roll top desks with numerous drawers, cupboards and pigeonholes to store the papers on which they worked. The end product was often large bound volumes containing hand-written records in support of the manufacturing and trading business. Much of this work was carried out on an individual basis, with little need for interaction between workers. In fact interaction, i.e. talking between workers, was positively discouraged in many workplaces. And so it remained for almost 50 years. Even the introduction of the typewriter and basic adding machines made little initial impact on the work processes involved.

The situation started to change in the early 1950s with the era of mass production and western society rebuilding itself following the Second World War. Machines were progressively introduced to undertake the highly repetitive and labour intensive tasks involved in text production, records management and process control. With the introduction of computers in the late 1950s and early 1960s, greater emphasis was given to the control and manipulation of data and the office worker's interaction with this central data store.

Progress from the 1950s to the present day has been extremely rapid, seeing developments in distributed data processing, personal computers, global trading and a fundamental shift away from manufacturing in favour of a service orientation of business. At the heart of these changes, the office worker becomes part of a 'knowledge-based' process rather than a processor of raw information, with this now being done by machines. The trend also saw the move from individuals working in isolation to working in groups, team working both locally, nationally and internationally. Organizations became 'project-based' where working groups were formed and reformed to deal with particular 'knowledge-based' tasks for a given project, after the completion of which the team was dismembered with individuals forming new groups to meet the needs of the next project.

Despite the dramatic change in the past 50 years, the buildings used to accommodate these functions have developed far less quickly. In fact, it is only since the late 1960s that we have seen significant improvements in the office environment provided to suit these new tasks and the equipment upon which they depend, although regrettably not all environments have seen improvements.

The modern office is a highly specialized environment, highly serviced and equipped with a range of goods from computers to furniture, and from fax machines to video conferencing. It demands the latest in communications technology and is the focus for innovation in many aspects of workplace design. Yet its history stretches back a little over a century.

Although there is mention of *the office* in classical and biblical literature, the contemporary office is the result of developments in the latter half of the 19th century. The work processes commonly referred to today as 'office work', and consequently the office, developed progressively out of the Industrial Revolution. Office buildings date from the 16th century. The Ufizzi in Florence, designed by Giorgio Vasariano built between 1560 and 1574, is probably the first purpose-built office – in fact 'Ufizzi' literally means offices.

However, there was no logical progression from the Renaissance model to that of the present day. For most of three centuries it was common practice to combine the home and work environments – a subject to which we will return in later chapters. Businesses were generally on a small scale employing few people, which for many revolved around family members, and hence the integration of work and home environments posed few difficulties. This self-contained environment was only superseded when improvements in communications enabled businesses to expand.

At the heart of the revolution in business methods was a series of inventions that were to change work and work processes irrevocably – in fact were to change the way we live. Of these, the three most significant were probably the telegraph, the typewriter and the telephone.

The earliest form of mass communication system was the telegraph. First introduced in 1844, it enabled people to communicate from one town to another and eventually all over the world. The size of a country like the USA meant that a telegraphic message could be relayed to many people thousands of miles apart within minutes. This advantage was not lost on business, where for the first time companies could separate administration from manufacturing, and establish branches in towns and cities across the country. The telephone, invented some 30 years later by Alexander Graham Bell played an even greater part in shaping businesses and their work environments, as people could use it not only to talk over great distances but as a primary means of internal communication between those working in the same building. Offices therefore soon developed as clearing houses for information. The advent of the typewriter, invented in 1866 by Latham Scholes, and first manufactured by Remington and Son, transformed the nature and pace of office work. Together, these three inventions enabled the office to develop into a centre of communications, allowing people to work together even if they were not in the same building – or even the same city.

The expansion of businesses such as insurance and mail order demanded a new type of working environment to accommodate the growing service industries. The model for many offices in the early 20th century was one that employed the most advanced aspects of office planning within a building specially designed for the purpose. The Larkin building in Buffalo, New York, designed by Frank Lloyd Wright in 1904 to house the Larkin Mail Order Company, accommodated 1800 personnel engaged in administrative and processing tasks. The workspace in the Larkin Building was designed to provide efficient but also humane conditions; however, its large open spaces gave workers very little personal privacy.

It is worth noting that during its early development the office environment was an almost exclusively male domain and 'office work' was regarded as a high-skill operation. Over the years office work progressively changed from typing and text production, through data input to increasingly complex processes involving several tasks. Rather than one repetitive task, workers progressed to carrying out several components of the work process and its interlinked activities. With the growth of business and the accompanying increase in the volume of work, the nature of office work began to change. Each transaction began to be broken into discrete tasks, such as typing a letter or making a telephone call, which parallels developments in industrial production where the manufacturing process was divided into its component tasks, each of which was the subject of individual analysis. It is not surprising therefore, that the layout of offices broadly mirrored factories where workplaces were arranged in serried rows with the manager or supervisor located such that they could oversee all of the process and workers for whom they were responsible.

Much of the development of office furniture and equipment was associated with what was perceived to be lower grade work, such as typing and computer operation where workers spend many hours in a fixed position in front of a typewriter or a computer screen. Executive furniture has always been self-consciously traditional, using conventional styles associated with power, authority and permanence. By way of example, despite the advanced design of the 1986 Lloyd's building, and what at the time was considered to be its unconventional and innovative layout of office space, the committee room (Board room) is decorated and furnished in 18th century Adam style.

As office work became gradually more regulated, scientific principles of organizational management were developed to achieve maximum efficiency. The speed of communications required that workers could process increasing amounts of information more quickly. New management structures were developed to allow supervision of the increased number of work groups performing different tasks. In 1925, W.H. Leffingwell wrote a book called *Office Management, Principles and Practice* in which he advised office managers to walk around and observe the posture of typists and clerks to determine the design failings of desks and chairs. No element of office life was too small to be overlooked; management was exhorted to work out the cost of time spent sharpening a pencil, and simple workplace settings allowed them to see work at a glance. Instruction on the use of the workplace and the evaluation of staff on their use of it and other tools, were commonplace, as this extract from a contemporary office manual indicates.

> Desk systems should be taught to all clerks and close watch kept until they have thoroughly learned them. To ascertain how well they are proceeding, suddenly ask for an eraser or a ruler, or some other item not in constant use, and see how long it takes to locate it. If it cannot be located at once, the lesson has not been learned.

So the office had developed as a 'paper factory' where people were cogs in the machine. The highly repetitive movements of typing where the individual was seated for hours at a time, required a seat that held the operator firmly in position. Chairs were developed that were made of wood, had a swivel base, a hinged back and could be adjusted in height. Although crude by contemporary standards, they were among the first examples of the practical application of ergonomic principles in the office. In the 1920s and 1930s metal furniture gradually replaced wood in the office, its appearance reinforcing notions of efficiency and cleanliness.

Construction of office buildings continued to increase after the Second World War and almost every piece of work-equipment underwent major changes. However, the office interior environment and its furnishings continued to reflect the hierarchy of the organization, with most interiors being rigidly rectilinear with enclosed workplaces sized exactly according to the rank of the worker. Although offices designed on this rational and analytical basis looked efficient, they did not necessarily produce the best results from those working within them. By the end of the 1950s a completely different approach to the organization of the office environment emerged in the form of a new movement, called *Bürolandschaft*, or 'office landscaping'. The concept was invented in Germany by the Schnelle brothers and was, for a period, adopted by office planners amid considerable controversy. The landscape was based on the idea of free-moving communications between different groups of workers within a large, open-planned space, which was a deliberate contrast to the idea that a rational and highly-structured working environment resulted in organized workers.

It was, some might say, a complete reaction to the rigid internal structures that were the hallmark of work environments throughout the preceding postwar decade. The 1960s, the age of self-expression, was manifesting itself in the 'free-form' environments of Bürolandschaft. Office landscaping eliminated barriers and revived the use of large, open floor space divided into manageable, interconnected areas. New materials such as carpets and curtains were introduced that were practical as well as attractive, as they absorbed sound. Plants not only acted as dividers but introduced a more relaxed and natural element into an otherwise predominantly mechanistic environment.

Bürolandschaft was opposed by many managers who believed that the apparently random arrangement of desks would produce chaos, but its justification was psychological. The visual informality and flexibility of a landscaped office was based on the idea that workers used to having distraction all around them are likely to become used to it and will therefore learn to ignore the interference of office life.

The landscaped office demanded a specific approach to workplace planning based upon informality and openness. However, these redeeming features could only be achieved at a price – that price was floor space. This and the other major drawbacks of lack of privacy and distraction, were soon to condemn this planning approach to the ranks of the 'also rans'. During the 1960s furniture was designed that was flexible enough to be mass-produced yet could provide a different design solution for every office. Herman Miller, a

(a)

(b)

Fig. 2.4 (a) and (b) The telephone and the personal computer transformed the nature and the place of work.

furniture manufacturer in the USA, produced its Action Office product range, which was a furniture system that incorporated its own partitioning in the form of self-supporting screens. The design was based on a number of interchangeable panels from which desk tops, shelves and storage units were hung. With Action Office the whole ethos of office furniture was changed: rather than providing pieces of furniture to fit a certain space, Action Office created the environment itself. Following the commercial success of Action Office, Herman Miller applied similar principles when it developed Action Factory for industrial and product assembly environments.

Desktop computers first began to appear in the office in the late 1970s with the developments of Apple and IBM's personal computer (PC). Soon this device was to transform not only our work lives, but all facets of our lives (see Fig. 2.4). These computers had profound implications for work, work processes and the future of the place of work. The prospect of workers, once more based at home communicating with personal computers, modems, faxes and phones, may as some predict, mark the beginning of the end of the office as a separate working environment (Veldon and Piepers, 1995).

At the same time, there has been a return to the idea that offices should be places of personal expression rather then rational uniformity. Contemporary office design uses a wide range of approaches to suit a variety of applications. In large offices, flexible systems combine efficiency with a consideration of human factors. The increasing dominance of the computer focuses on new considerations for the health and comfort of the worker.

Ergonomic and environmental issues in the office are as much of a challenge today as they were at the beginning of the century (see Table 2.3, Fig. 2.5). As we move towards the advent of the paperless office, the focus will turn more and more towards the comfort and effectiveness of the user and the role that the workplace plays in the performance of the worker.

The last 20 years has seen a rapid increase in competitive pressures across all spheres of business. That competition, brought about in part by the freeing up of markets and in part by increasing global competition, has led to great turbulence and uncertainty for most organizations. Once upon a time, organizations could plan 3–5 years ahead with some confidence in all aspects of their business and the commitment to a lease or the purchase

Table 2.3 The history of the office via technology

1844	Invention of the telegraph – communication of distance
1854	Invention of the elevator (lift) – high-rise buildings
1860s	Steel frame and curtain walls developed – higher-rise buildings, the skyscraper
1866	Invention of the typewriter – mechanization of text production
1876	Invention of the telephone – ease of communications over distance
1901	Invention of punch keyboard – fore-runner of the computer
1930s	Window air-conditioning developed – deeper floors
1950s	Centralized air-conditioning developed – even deeper floors and suspended ceilings
1970s	Energy crisis - energy conservation research and development, and building automation systems
1970s	Mainframe computers are in wide use – centralized, fast data processing
1977	Apple launch the first microcomputer – distributed, even faster data processing
1979	Ericsson launch the mobile phone – ease of communications from anywhere to anywhere
1980	IBM launch the Personal Computer – ubiquitous data processing
1985	Microsoft launch Windows – expansion of easy to use, interrelated applications
1987	Ubiquitous PCs – bringing with them LANs, WANs and cable management problems
1990s	Portable technology abounds – mobile phones and computers enable work to be done anywhere
2000s	???

Fig. 2.5 The Lloyd's building in London, flexible and technically highly sophisticated workspace.

of a building was seen as an investment. This was when the demand for property was far greater than the supply, and hence gave rise to increases in price.

The 1980s was a time of correction. Continually to compete in the face of increased competition, we saw 'downsizing' or 'right-sizing' as companies slimmed down, leaving buildings empty with many years of the leases remaining. In parallel, organizations came to terms with the fact that they simply could not plan ahead more than one or two years into the future. However, what most organizations failed to understand is the positive or negative contribution that the workplace could make to corporate performance.

The 1980s also saw the emergence of call centres as a new work setting, where the combined use of telecommunications and data systems changed the face of many businesses' sales and customer service operations (see Fig. 2.6). Although intended as an efficient way of centralizing telephone contact with customers, the environment of many centres is that of a white-collar factory with row upon row of identical desks staffed with teleworkers who are monitored so closely that even their toilet breaks are timed, giving rise to them earning the label of 'the sweatshops of the 1990s'.

The rapid growth of call centres through the 1990s is forecast to continue with some analysts estimating that the current 1% of the UK's working population employed in this way, will double by 2001 and rise to 10% or more by 2005.

In many call centres, the footprint of the agent's workplace has been reduced to 2.5 sqm or less, and overall workspace densities of 8 sqm per person are commonplace. Add to this the high expectations the business has in terms of target achievement and the often stressful situations call centre agents have to handle, then it is not surprising that businesses operating with such environments demonstrate staff attrition rates of over 30% per annum. Consequently, for the call centre to be an effective business tool, businesses need to provide environments that offer 'compensation' for the nature of the job rather than

(a)

(b)

Fig. 2.6 (a) Call centre, Orange, Darlington; (b) Subscriber Management Centre, Sky, Dunfermline.

exacerbating the pressures. However, as we will discuss later, it is possible that the call centre as a place of work will reach its zenith between 2005 to 2010, to be followed by a rapid decline to oblivion over the subsequent decade.

Industrial or commercial, private or public sector, large or small business, whatever the job, wherever the location, the effective workplace must be seen as more than just a *place* (it is a combination of three components – the physical work setting, the information environment and the service environment) required to support the individual in his or her endeavours. In the chapters that follow, we will describe how this integrated workplace can be created, serviced and adapted to the changing needs of business.

References

Balch, W.F. (1994) An integrated approach to property and facilities management. *Facilities*, **12** (1), 17–22.

Becker, F. and Steel, F. (1995) *Workplace by Design – Mapping the High-Performance Workscape* (San Francisco: Josey-Bass Publishers).

Bon, R. (1989) *Building as an Economic Process* (Englewood Cliffs, New Jersey: Prentice Hall).

Brandt, R. (1993) What is strategic facilities planning? *FM Journal*. IFMA. May/June, 36–39.

Byrne, J.A. (1992) Management's new gurus. *Business Week*, 31 August, pp. 44–52.

Champy, J. and Hammer, M. (1993) *Re-engineering the Corporation – A Manifesto for Business Revolution* (London: Nicholas Brealey) p.17.

Crainer, S. (1996) That was the idea that was. *Management Today*. May, 38-42.

Duffy, F. and Tanis, J. (1993) A vision of the new workplace. Site Selection, 38(2) Industrial Development Section 427–432.

Duffy, F., Laing, A. and Crisp, V. (1993) *The Responsible Workplace – The Redesign of Work and Offices*. (Oxford: Butterworth Architecture, in association with Estate Gazette).

Eley, J. and Marmot, A. (1995) *Understanding Offices* (London: Penguin Books).

Flanagan, R., Norman, G., Meadows, J. and Robinson, G. (1989) *Life Cycle Costing – Theory and Practice* (London: BSP Professional Books).

Holmes, R. (1988) A systematic approach to property condition assessment. In: Then, D.S.S. (ed.) *Whole-Life Property Asset Management* (Heriot-Watt University), pp. 140–150.

Kimmel, P.S. (1993) The changing role of the strategic facilities plan. *FM Journal*, IFMA, May/June, 24–29.

Marshall, H.E. and Ruegg, R.T. (1990) *Building Economics – Theory and Practice* (New York: Van Nostrand).

Parshall, S. and Kelly, K. (1993) Creating strategic advantage. *Premises and Facilities*. August, 11–16.

Sager, I. (1995) Big Blue's white-elephant sale. *Business Week*, 20 February, 36.

Spedding, A. (ed.) (1987) *Building Maintenance, Economics and Management*. (London: Spon).

Then, D.S.S. (1994a) Asset management and maintenance – Part 1. *Facilities Management* **2** (2), 10–12.

Then, D.S.S. (1994b) Facilities management – the relationship between business and property. *Proceedings of EuroFM/IFMA Conference on Facility Management European Opportunities*, Brussels, p.259.

Then, D.S.S. (1994c) Property as an enabling resource to business. *Facilities Management*, **1** (6), 5.

Then, D.S.S. (1994d) Property as an enabling resource to business. *Facilities Management* **1** (6), 4–7.

Then, D.S.S. (1994e) People, Property and Technology – managing the interface. *Facilities Management*, **2** (1), 6–8.

Then, D.S.S. (1995) Asset management and maintenance – Part 2. *Facilities Management*. **2** (3), 7–9.

Then, D.S.S. and Fari, A. (1992) A framework for defining facilities education. In Barrett, P (ed.), *Facilities Management – Research Direction*. University of Salford. 15–22.

Varcoe, B. (1996) A key to unlock the future. *Facilities Management*, **3** (5), 6–8.

Veldon, E. and Piepers, B. (1995) *The Demise of the Office* (Rotterdam: Uitgeverij).

3

The space planning process

In the previous chapters we have defined space management and made the case for businesses to plan their workspace in terms of its type, amount, attributes, location and quantity, to meet business needs through time. In this chapter we set out to describe the main components that a space planning process should possess in order that it successfully delivers the objectives of the business.

At the outset it should be recognized that there is no one definitive method that should be adopted universally by all organizations for the planning of workspace, since each will be shaped by factors such as:

- The relative maturity of the organization in terms of space management;
- The resources and tools it has at its disposal; and, most importantly of all,
- The dynamics in its marketplace and how it needs to respond to them.

What follows therefore is a description of the key components of planning processes that we have encountered and applied in our work with many organizations.

3.1 The need for an accommodation strategy

Just as the aim of a business strategy is to provide a framework that guides the organization in its decision making towards a chosen business goal, similarly an accommodation strategy provides the framework for decision making relating to the management of an organization's workspace.

Unfortunately, all too often the fact that an organization evolves rapidly, whether it be by organic means or by acquisitions and mergers, is seen as justification for not producing a strategic accommodation plan. When questioned, facilities managers in such organizations typically respond by saying that even if they had time to develop a plan, which due to pressure of work they do not, the plan would be meaningless because the business is changing so quickly the plan would be out-of-date as soon as it was prepared. While we recognize the temptation in a rapidly developing business is to say 'as we are changing so fast it is impossible to plan on any firm base, so we will not plan at all', we believe that of all organizations, it is these who need and can benefit most from developing a plan. If the organization is growing or contracting fast it is even more important that it has a plan of how its workspace needs can be met through time. The results all too often experienced as direct consequences of the failure to plan, can typically include:

- Insufficient available workspace;
- Inappropriate premises;
- Inefficient facilities and services;
- Outmoded systems for handling information;
- Conflict between facilities providers and facilities users;
- Poor staff morale; resulting in
- The achievement of poorer than expected productivity levels.

For success therefore, there is a need to consider these and other issues at an early stage and, in so doing, create a framework around which the organization's workspace can be developed to function effectively. The production of a plan fosters an atmosphere in which decisions relating to the business's workspace can be made, confident in the knowledge of being able to respond to the varying needs of the organization through time. When well constructed, the plan not only enables responsive actions to be initiated, but will provide a basis for anticipatory actions by the facilities manager. This can be said to be the challenge for all workspace planning – to move from a reactive to an anticipatory process.

The Workplace Best Practice Group[1], a UK-based organization who conduct research and corporate benchmarking activities, developed the following definition of a strategic accommodation planning process for their work in this area, which we consider encapsulates all that is relevant:

> A strategic accommodation planning process is the method used by the organization to ensure that their Facilities, Infrastructure and Services are capable of meeting the seen and unforeseen needs of the organization, in order that they are able to cost-effectively sustain changes in their external business environment.

The strategic accommodation plan therefore has its roots in the organization's business plan. Property as a key asset of any organization and as a primary facility in support of the people in the organization, must be reviewed, selected and operated in an efficient manner which addresses the needs of the business. This cannot be done effectively without the aid of an appropriate plan.

It is important to appreciate that, to be effective, the accommodation plan needs to be capable of operation at both strategic and tactical levels. Further, the process employed for developing the plan must be founded upon a clear understanding that the plan's successful operation is dependent upon the effective inter-linking of the strategic and tactical components, as indicated in Fig. 3.1. Time and again, we visit organizations who have developed one, or sometimes both aspects, yet through their actions demonstrate their failure to appreciate that strategic and tactical planning are integral parts of the same whole, each responding to, and being influenced by, the other.

The reasons for this 'missing link' can be manifold, frequently however, it is evidence of the way the organization and its workspace is managed. The responsibility for strategic planning often falls within the domain of real estate and property managers while tactical planning is usually the responsibility of facilities managers. While undoubtedly the skills

[1] The Workplace Best Practice Group is a club based in the UK which is facilitated by AWA Best Practice Benchmarking Ltd, London.

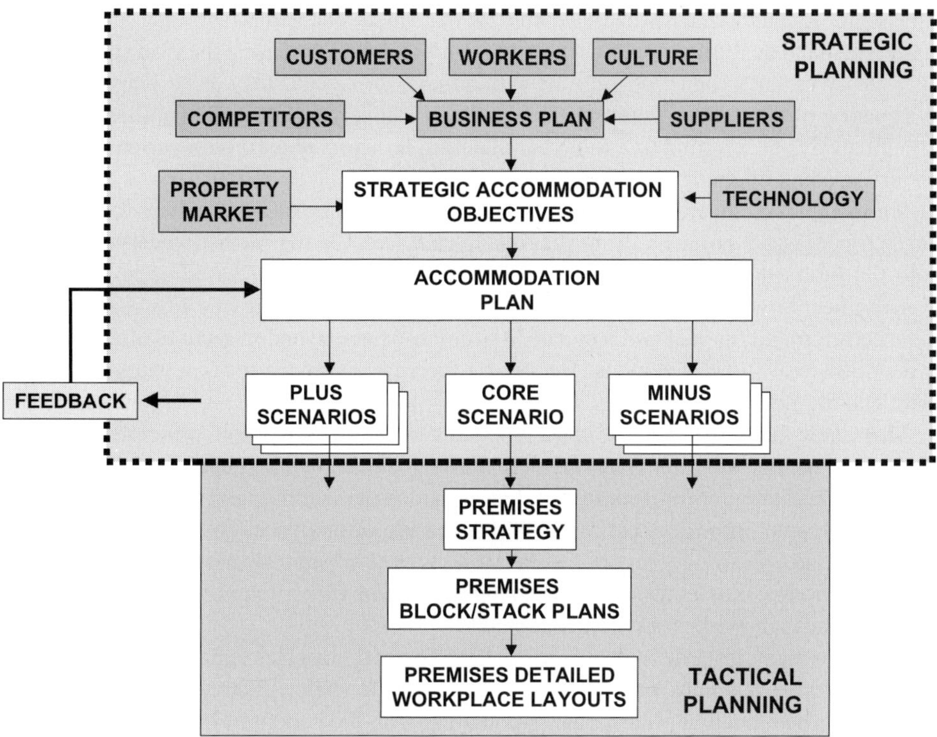

Fig. 3.1 The elements of an accommodation planning model.

required for each aspect are different and may well be best provided by different people depending upon the way the organization is structured, there can be no justification for the separation of such vital and interdependent activities.

As can been seen in Fig. 3.1, there are sequential steps which need to be taken, leading from the enterprise's business plan to the provision of an individual workplace. Although, as will be discussed later, there is a high level of interaction between each stage, the steps need to be taken in their logical sequence. Each will require information upon which decisions will be based, and many will share the same data. However, that too poses a challenge for those engaged in accommodation planning. It is important for them to distinguish between the level of information required at the relevant stages in the process so as to avoid being encumbered by the sheer weight of data that will need to be gathered over the course of the whole accommodation planning process.

3.2 Strategic planning

Any good business plan will attempt to address the objectives of the organization over the short, medium and long terms. In this context, short is considered to be up to five years, medium to be 5–10 years and long term more than 10 years.

For many organizations, the prospect of preparing a five year facilities and accommodation plan will be a new departure. For some it will be considered unreasonable and without any valid foundation. However, when one considers the length of time involved in gathering data, analysing it, preparing a plan and then implementing the plan (particularly if it involves the production of a new building), 2–3 years may have elapsed before occupation of the building can even begin. It is for this reason that the facilities manager should always be working to a five year planning horizon, albeit it with various scenarios to take account of changes that may arise in the intervening period.

We recommend that an approach we call 'the rolling 5' be adopted, where forecasts are made for each year going out from year one to year five. Clearly, as the forecasts go further into the future then they will be less precise and most probably based upon a range of possibilities, in which each year is wider than the preceding year (see Fig. 3.2). Then as time comes round for year two's update, it becomes year one and so on until after year four a new fifth year forecast is made. Hence, the forecasts are made for five years at any one time and are rolled forward.

Most new buildings are designed and constructed based upon principles that are predicated on a 60 year life. Most UK institutional leases are based upon a 25 year period. In either case, long term planning is necessary since the commitments occupying organizations will enter into for either ownership or leasing of premises will, by most standards, be considered to be long term obligations. If no long term strategic planning of labour and space requirements is made, then the foundation upon which long term accommodation commitments are made, must be questionable.

In far too many instances, organizations have found themselves suffering in the medium term (let alone the long term) through the failure to plan effectively. This is not the comment of an armchair philosopher expounding with the benefit of hindsight. We can all be wise after the event. Rather these comments are made to encourage those involved in long term planning to maintain their vision on that distant planning horizon, as a backdrop against which the needs of 'today and tomorrow' can realistically be made. Where financial and/or practical considerations prevent commitments being made now to sites and buildings to meet the needs of 10–20 years ahead then it is not sufficient just to ignore their implications. Long term strategic planning does not necessarily involve buying that site, or acquiring that building, but rather considering 'what if' each time a decision is being taken to test its validity against the long term strategic plan. For example, when

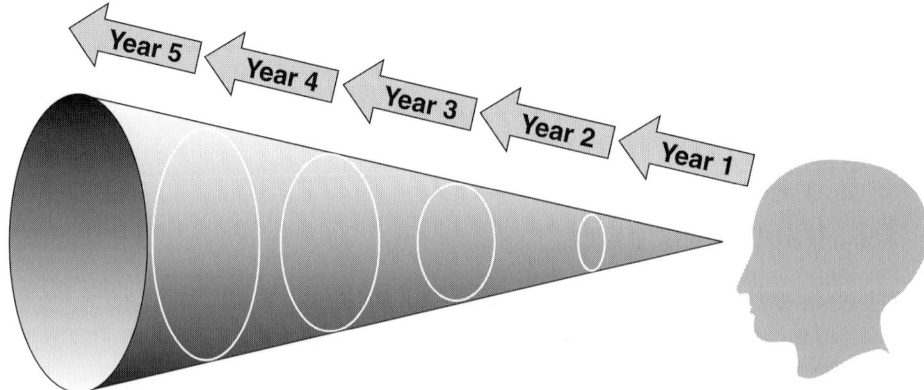

Fig. 3.2 Progressively over a 5 year period forecasts become narrower in range and hence more definitive.

committing to new premises (whether purchased or leased) consideration should be given to 'what if' the organization contracts within the next 5–10 years, will the premises be easily let or sold? What if the organization expands, how easy will it be to extend the premises, or acquire the adjoining site or building? Consideration of such matters at the time when the decision is being taken to meet the needs for 5 years hence, can save much corporate pain, both physical and financial, in the future.

The need to develop new forms of accommodation strategy is recognized by Franklin Becker when he says 'It takes a fundamental paradigm shift to identify and implement workplace strategies that alleviate the pressures organizations are facing as they struggle to become more competitive' (Becker and Steele, 1995, p. 103 and p. 109). He then goes on to outline how the basis of accommodation strategies needs to change from cost-driven models to business driven models, as organizations strive for improvements in organizational effectiveness and profitability (Becker and Steele, p. 123).

This is where the linkages between the business and accommodation plans are made, and where strategic and tactical plans play a vital role in the successful application of the accommodation forecasts. Scenario models are developed in response to various possible situations the business may encounter. Typically, the information required to develop scenarios would involve:

- Organizational and business structures and groupings.
- Numbers of people required to deliver the business objectives – employees and others.
- The work processes required to deliver the goods or services to the business's customers.
- The application of technology to the channels through which the organization delivers its business.
- Locational criteria in relation to customers, suppliers and competitors.
- Sales and transactions forecast by business units.

From this information a *core scenario* can be developed which is the organization's best view of the future. Best in this context is not 'most optimistic' but is the 'most reliable' forecast based on a full understanding of the business environment in which the organization operates. From this core scenario, other possible scenarios are developed, by applying sensitivity analysis techniques which have the effect of acting as pluses or minuses in relation to the core scenario. The criteria to be applied will be specific to each organization in the context of its business operations; however, the following are indicative of the types of factors that could be applied.

- The impact of different levels of sales and transactions, their sensitivities and dependencies.
- The relative pace of product development.
- The impact of social, economic, political and legislative changes.
- The impact of climatic and environmental changes.
- The significance of demographic changes.

From these 'alternative worlds' of business, several corresponding accommodation scenarios can be developed to meet the needs of different levels of business, different numbers of people in the organization, different levels of technology usage, different organization structures and so on.

The accommodation planning process described here and illustrated in Fig. 3.3 comprises two discrete but inter-linked sub-processes, i.e. strategic and tactical components

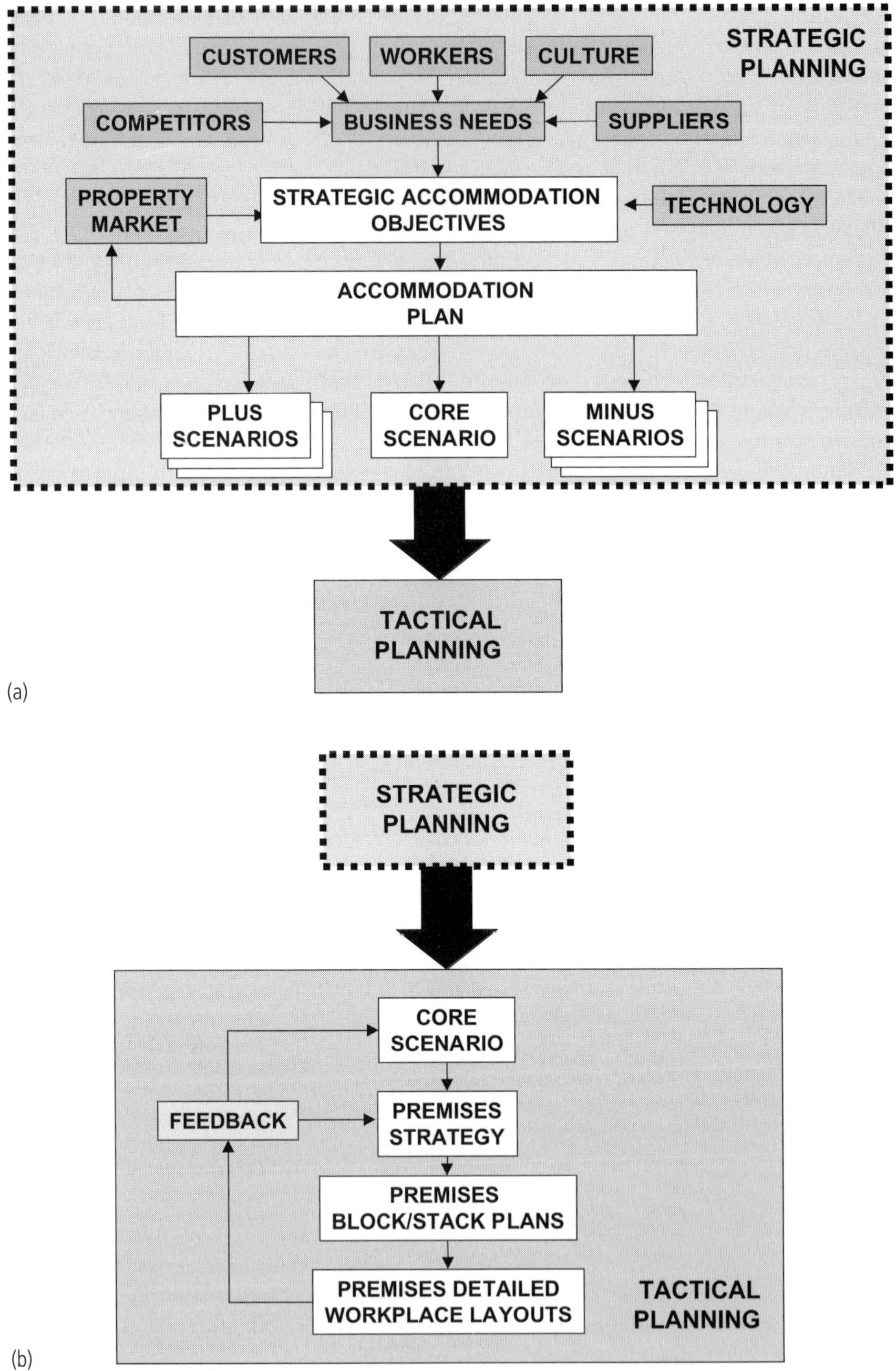

Fig. 3.3 (a) and (b) In developing a comprehensive accommodation plan, the output from the strategic planning process is the input to the tactical planning process.

(see Fig. 3.4). Each sub-process can be operated independently in response to planned initiatives and reactive triggers and each will have its own cycles, timescales, databases and information 'shelf-lives'. However, within the operating cycle of each sub-process there need to be triggers which activate a review and possible updating of the other sub-process.

The accommodation planning process described here follows the conventions of: investigation and research, analysis, development of solution(s), implementation, review and feedback. There are several publications that document how to develop a brief and capture the data upon which it is based, to which we can add nothing but encouragement to readers to consult (O'Reilly, 1987; Barrett, 1995).

3.2.1 The strategic accommodation plan

The accommodation plan is itself a component of the organization's facilities plan. The objective of the strategic plan is the creation of a vision for how the accommodation requirements of the organization can be satisfied to meet different and varying business circumstances. Consequently, to develop a strategic plan it is necessary to capture, at a high level, information about the organization, its work processes, its people and its existing accommodation and how it is anticipated that these may change in the future.

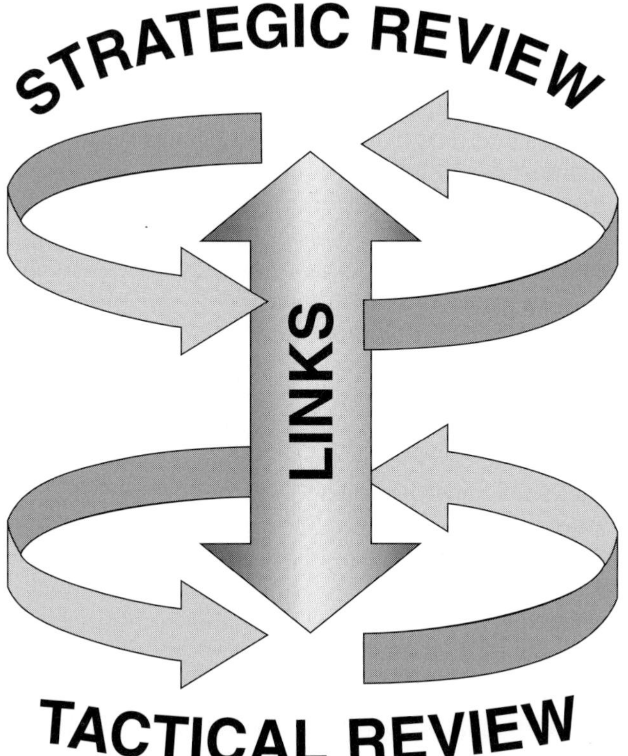

Fig. 3.4 It is essential that the review processes of both the strategic and tactical plans are connected to enable a two-way exchange of data and feedback.

Most of the elements of a strategic accommodation plan are required by the business, not just for the sake of developing the plan, but for the effective and efficient management of the business itself. This should, but often does not, make the process of capturing the relevant data easier than it might otherwise be. The avoidable difficulties result from a failure by the business to appreciate the need for sharing the required information and its real value to the enterprise as a whole. This is a subject to which we will return later in this chapter.

Developing the strategic accommodation plan will involve many activities following a logical and predetermined process. In this chapter we outline the process and the principal headings under which the organization's needs for property, and the property itself, are assessed.

The process proposed by the Workplace Best Practice Group, for preparing an accommodation strategy, involves the following ten steps.

Step 1. Understand the property portfolio; this requires a rigorous and thorough appraisal of the current property portfolio, its infrastructure and the facilities provided, leading to the compilation of a database that will form a key tool in the planning process.

Step 2. Understand the organization and its work processes, possibly more comprehensively than has been the case in the past, in order to establish the relationship between the workspace and services that are provided.

Step 3. Understand in detail the activities within each work process and their drivers for space.

Step 4. Understand how the organization and work processes map onto the current property; this involves making the link between the way that the business works and the space and services required by the different elements of the organization.

Step 5. Understand the external influences that act upon the business at a macro level, such as its customers and suppliers, as well as the external social, economic and political environments in which it operates.

Step 6. Understand the likely business scenarios for 1, 2–3, 3–5 years ahead, which is the creation of a series of 'what if' situations.

Step 7. Model the implications for the organization, its work processes and property, which each of these scenarios might have, charting the changes that may be required to the business's property portfolio.

Step 8. Understand the implications for the workspace drivers, by linking these scenarios to the need for different types of space, facilities and services.

Step 9. Through intensive modelling, show how the existing property can meet the defined space needs, identifying changes that are necessary in terms of adaptation, acquisition and disposal, and under what circumstances each would be required.

Step 10. Develop migration plans, which indicate how the workspace will evolve from the current state to the future requirement.

Another misconception often encountered relates to the term itself, i.e. strategic plan. In the singular form it can lead to the belief that the outcome of the process is a definitive plan that will set out the organization's needs and how these will be met. At one level of interpretation this is correct; however, to enable the plan truly to meet the needs of the business it must be capable of responding to changes that the business will experience.

Hence to do so, without the need to re-draft the plan on each occurrence, the plan needs to comprise a series of sub-plans, each of which has been developed in response to different possible business scenarios. This is where the dependence upon, and interaction with, the corporate business plan comes in.

One of the principal reasons why, even those who attempt to prepare a strategic accommodation plan, do not meet their objective, is that the facilities plan, which includes details of the accommodation, environmental services, information technology systems and the like, is prepared and coordinated independently of the business plan. Ironically, it is the effects of the business plan that create the needs which have to be addressed by the facilities plan, yet seldom is the facilities manager involved in developing the business plan that has such a dramatic effect on his or her role, tasks and responsibilities.

It is therefore important to tie the business planning process together with the facilities planning process, to ensure comprehensive integrated planning. When this is done, the facilities required to support the organization, including accommodation, can be forecast and budget provisions made, to ensure that facilities of the right type and quantity are available where and when required.

3.2.2 The tactical accommodation plan

As has been explained, the start point for tactical planning is the output from the strategic plan, which defines the overall framework within which workspace can be created in specific buildings. Different scenarios will have been modelled to assess the capabilities of these premises to address the possible changing workspace needs of the business through time. For the selected scenario the facilities manager now embarks upon the process of developing a workplace layout for each person involved in the business's operations based upon a more detailed understanding of their requirements. Some of the information will have been gathered to enable the strategic plan to be developed, which now needs to be considered in more detail. To economize on effort, it may be considered appropriate to collect both levels of information at the same time, depending upon the context of the exercise as a whole.

The central aim of the tactical plan is to provide a framework for workplace layouts that will meet the needs of every work group in the business, today and over the short-term. Where the total cost of occupancy of workspace can amount to as much as 10% of turnover for some businesses, then getting the balance right between meeting the needs of today and tomorrow can have a significant impact upon the organization's financial bottom line.

3.3 Stages of the planning process

The first part stage in developing an accommodation plan is best considered in two parts as it involves gathering data about two quite different aspects, namely the organization and the buildings to be occupied (see Fig. 3.5). Most frequently, the sources of the information under each of these headings will involve different people in different locations, and where the buildings are not under the control of the organization (e.g. a building new to the organization) then possibly it will also involve third parties.

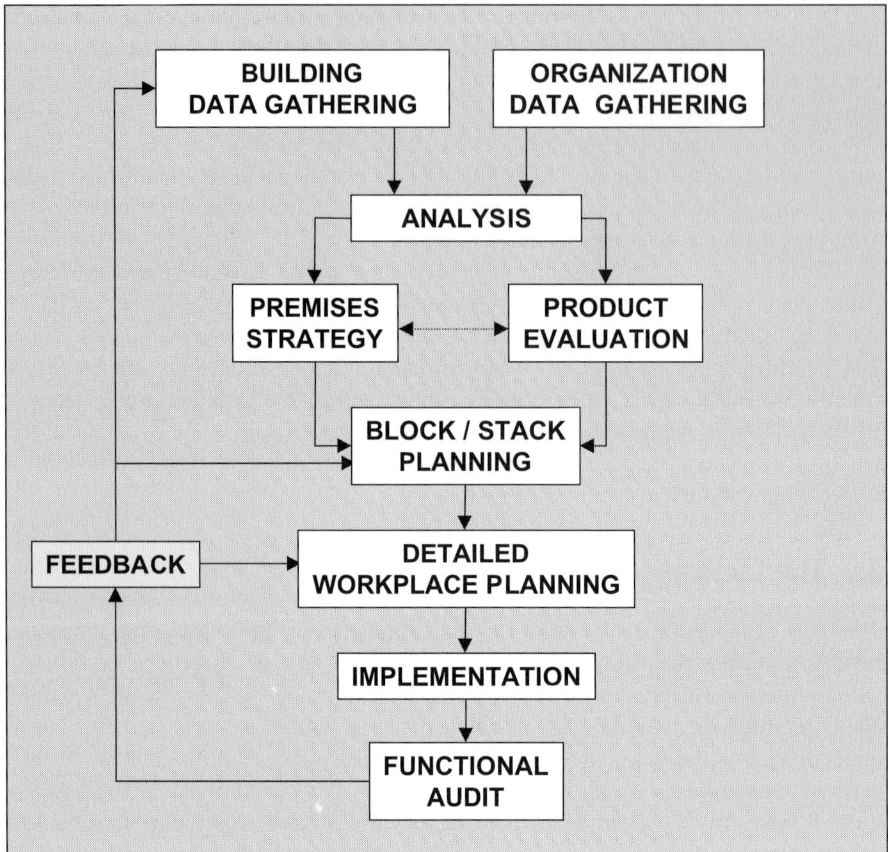

Fig. 3.5 The elements of the tactical planning process.

3.3.1 Gathering data about the organization

This stage involves compiling information about the organization to be accommodated, its culture and experiences, how it works, its needs, and how these may change in the future. The methods used to collect this information will be dependent upon the extent and currency of existing data, the tools employed and the resources available within the organization. The most frequently used methods are as follows.

- **Customer interviews** – with key personnel representing each function and/or work process of the organization's operations, and who will be occupants or users of the accommodation. The interviews should be conducted using standard questionnaire forms, to ensure that the process is consistently thorough, detailed, systematic and objective. The principal information gathered will relate to organizational structure and groupings, job functions, work processes and working practices, personnel forecasts, relationships and communication patterns, particularly those that are face-to-face.
- **Specialist interviews** – with key specialists involved in supporting the organization's existing operations to obtain detailed information regarding aspects such as use of

information technology and communications systems, document retention procedures and storage methods, health and safety, facilities services and their provision.

- **Workshops** – to be attended by a sample of personnel representing the relevant parts of the organization, and which are facilitated to encourage the exploration, in a highly interactive way, of work and work processes, the workplace, its servicing and related issues. The aim is to simulate a typical and atypical working day in the organization and in so doing consider holistically the key elements necessary to support work effectively and how they relate to each other.
- **Focus groups** – interactive sessions attended by a small group (up to 8) people who have a detailed knowledge and/or first-hand experience of the topics under consideration, which are examined and discussed in detail, with a view to obtaining quick, reliable, high quality input/feedback.
- **Workplace Effectiveness Appraisal Questionnaires** – submitted to a sample of personnel representing the relevant parts of the organization to assess the customers' views on the effectiveness of the existing workplaces in use, and how well they support them in carrying out their jobs, especially in solitary and team working activities.
- **Working Practices Questionnaires** – submitted to a representative sample of personnel in the organization to capture details of the work tasks they carry out, their location, duration and frequency.
- **Utilization Surveys** – involves measuring the levels of occupation of existing workplaces throughout the working day or week, typically conducted over a five week period. The extent of the surveys will be dependent upon the circumstances prevailing in the organization and may range from hourly to four times per day, i.e. two in the morning and two in the afternoon.
- **Observation** – of the existing workplaces in use, the pattern of activities carried out, the interactions and the flow of people through the building, and their use of the facilities and services provided.

Using these techniques, or a combination of them, should enable sufficient and appropriate information to be obtained to develop a detailed 'picture' of the organization, its people and its work activities. All of the information gathered needs to relate to the existing business needs and how they are likely to change in the future and over what timescale. Of crucial importance is a 'people centred' approach to the data gathering and briefing process, as advocated by Marilyn Standley who employed just such a highly consultative approach in her role as project director for the development of the new headquarters for Addison-Wesley Longman Group at Harlow in Essex, UK (Edwards, 1998).

Several of the methods described can be used to capture similar data. While excessive duplication of data gathering should be avoided, benefit in terms of reliability of facts can be derived from the cross-checking of data collected via different methods. For example, although the working practices questionnaires will, amongst other issues, elicit information about the amount of time people spend on particular tasks, data gathered from utilization surveys have often been found to be more comprehensive and hence more reliable.

Having collected data about the organization, next, information is required about the premises that the organization occupies or is considering occupying.

3.3.2 Gathering data about the premises

The objective of this stage is to develop as comprehensive an understanding of the business's property portfolio as is possible. This will involve gathering information about the premises to be occupied, their condition, infrastructure, tenure, past and present uses all with a view to understanding the capabilities of the building to respond to the needs of business.

As with the organizational data, the methods used to collect building information will be dependent upon existing records, how up-to-date they are and their completeness. The most frequently used methods comprise the following.

- **Specialist interviews** – with key specialists involved in the building's operations, maintenance and management with a view to obtaining information regarding aspects such as condition, running costs, environmental services, planning constraints, infrastructure, history of alterations, tenure and legal obligations.
- **Focus groups** – attended by personnel representing building users as well as specialists involved in the running of the building, to consider how well the premises and infrastructure perform in support of the needs of the organization. The advantage of such a constituted focus group is that it is able to consider aspects of the building's operations individually as well as holistically, from the standpoints of both 'suppliers' and 'customers' and, in so doing, gain a better understanding of the effect these operations have in supporting the people in the business and their work.
- **Workplace effectiveness appraisal questionnaires** – where the building is currently occupied by the organization, then the previously referred to questionnaire used to capture organizational data can be extended to assess the users' views on the effectiveness of the building, its infrastructure, services and management – refer to Fig. 3.8.
- **Physical surveys** – involving the measurement of existing workplaces and support facilities, their condition and an assessment of the completeness of existing building and workplace record drawings. Where reliable record drawings are not available, then a full measured and condition survey of the building will be required.
- **Workplace assessments** – which need to be carried out for compliance with health and safety legislation may be available for review. If not, then these assessments would need to be conducted, which can provide useful information on the suitability of each existing workplace for the tasks to be performed and the comfort of the user.

The volume of data to be gathered about the organization and its buildings may at times appear awesome. However, this should not be used as a reason for cutting corners, since the quality of the accommodation plan to be prepared, is predicated upon the quality (accuracy and completeness) of this information.

3.3.3 Analysis

The information that has been gathered needs to be coordinated, collated and referenced so that it is in a usable form and its sources can be traced should any queries arise that need to be investigated further.

The nature of the analysis will be specific to each planning exercise; however, the aim of this stage is to convert the collected data into a comprehensive picture of the organization and its buildings, upon which the creation of the workplace environment can be based. As previously stated, the information gathered should relate to the current needs and operations of the organization and the best forecast that it is possible to achieve for a minimum of five years ahead. Typically we would expect the analysis stage to comprise the following elements.

Organization

- **Objectives** – for the organization and the businesses within it and their vision of its future development, cultural values, customer and staff expectations.
- **Structure** – current, recent past history and known future developments, including business, functional and process groupings.
- **Personnel** – numbers, type (i.e. employee, sub-contractor, agency etc), job and task descriptions, health and safety policies and procedures.
- **Adjacencies** – a map of the interactions between each functional and work process group within the organization, and also with third parties, their inter-dependencies and relationship modelling.
- **Work settings** – detailed descriptions of the 'hardware' requirements for each job function and process, space standards, generic workplace 'footprints' and operating guidelines (see Fig. 3.6).
- **Work practices** – communications, work patterns and timings, allocation and use of workplaces, correlation between work activities and the use of space.

Fig. 3.6 A detailed description of the 'hardware' requirements of job function and process are required – Call centre cluster (McQueen, Pentland Gait, Edinburgh).

- **Knowledge management** – information sources, access and retention policies, storage methods and location, service and delivery needs and security.
- **Quality** – amenity facilities, life support requirements, environmental standards, review procedures, communications and the 'fun' quotient.

Buildings

- **Location** – in relation to transportation nodes, staff and other labour resources, customers, suppliers, competitors, other parts of the organization and external services needed by staff.
- **Image** – in relation to customers, staff, competitors, the property market and the community at large.
- **Profile** – shape, building widths, heights, structural grids.
- **Areas** – gross internal, net internal, circulation, usable and wasted space.
- **Circulation** – fire escape routes, fixed corridors, stairs, lifts (passenger and goods), and escalators.
- **Environmental services** – HVAC, electrical power, lighting (daylight and artificial), location, condition, scope for relocating, controls and capacity.
- **IT services** – voice and data distribution, outlet presentations, condition, scope for relocating and capacity.
- **Planning** – the opportunities and constraints for the creation of different types of task settings, degrees of enclosure of workplaces, layout options, workplace capacity, services distribution and extendibility.
- **Uses** – current and past patterns of use, workplace types and locations, ancillary spaces, common support spaces, process environments, storage facilities, amenity spaces, and health and safety compliance.
- **External environment** – access and egress arrangements, car parking, security, amenities, neighbours, development opportunities and restrictions, and known future developments.
- **Running costs** – for the provision of the building, its infrastructure and its services.
- **Legislation** – the history of building regulations, environmental restrictions, planning restrictions, health and safety, and building uses.
- **Tenure** – lease terms, ownership restrictions, development restrictions, value, disposal opportunities and constraints.

The output from the analysis stage will be a Statement of Requirements (SoR) setting-out in great detail the organization's needs in terms of workspace and how these needs may change through time. We recommend that each customer group (department, work process group, project team etc) in the organization, is given a copy of the analysed organizational data that relates to their group, which not only keeps them informed by way of feedback from the data gathering stage, but can be used to verify the data and the conclusions that have been drawn.

Analysing information in a vacuum is of very limited value, as it is all too easy to focus on 'doing things right' at the expense of 'doing the right things'. The use of benchmarking techniques can be especially beneficial in this regard as they can enable a context to be set for the analysis of the organization's requirements. However, to be of value, benchmarking assessments need to be undertaken using directly comparable data, which require consistency in the data structure and in the terms and definitions used. For example, the Workplace Best Practice Group prepares extensive standards documentation for each of the study topics to be investigated, setting out terminology, classification of workspace, and detailing aspects expressly included and excluded from study. Having done this, comparative analysis of members' data can be carried out confident in the knowledge that it is being done on a 'like-for-like' basis, in sectors, cross-sector and 'best in class', as well as against a best conceivable practice profile.

3.3.4 Development of a premises planning strategy

The strategy is developed from a clear understanding of the capabilities of the building together with a statement of the organization's requirements previously identified through the analysis stage. This use strategy will detail what can and cannot be achieved within the building having taken account of corporate policy, tenure restrictions, cost and technical practicalities.

The objective of the premises planning strategy is to set-out how the organization's buildings can best be used having considered all of its capabilities (see Fig. 3.7). It will define the opportunities that exist for different workplace configurations and uses, and the constraints within which the business will need to operate while in occupation of the premises. Clearly, if the building's constraints impinge upon the organization and its work requirements to too great an extent, then this will need to be addressed either by alteration of the building or its disposal in favour of alternative premises.

Although each premises' strategy will be specific to the building and its intended occupants, we would expect all strategies to define the following as a minimum:

- **Workplace capacity** – a definition of the capacity of the building, in terms of the finite number of workplaces of each type required by the organization, which could be accommodated in the premises, and the space efficiency of such provisions.
- **Occupant capacity** – a definition of the capacity of the building, in terms of the finite number of occupants that could be accommodated in the premises, under different styles of working practice and within the constraints of area, life support services and legislative compliance.
- **IT and environmental services** – the location and distribution of services and plant, their presentation in the workspace, their capacity and flexibility to cope with reconfiguration.
- **Circulation** – the constraints imposed in response to legislative, health and safety compliance, vertical (lifts, escalators and stairs) and horizontal routes, and the flexibility that exists for reconfiguration.
- **Division** – the scope that exists for dividing the workspace into various types of enclosure to suit the requirements of workplaces, common support spaces, process environments and other accommodation needs of the occupants.

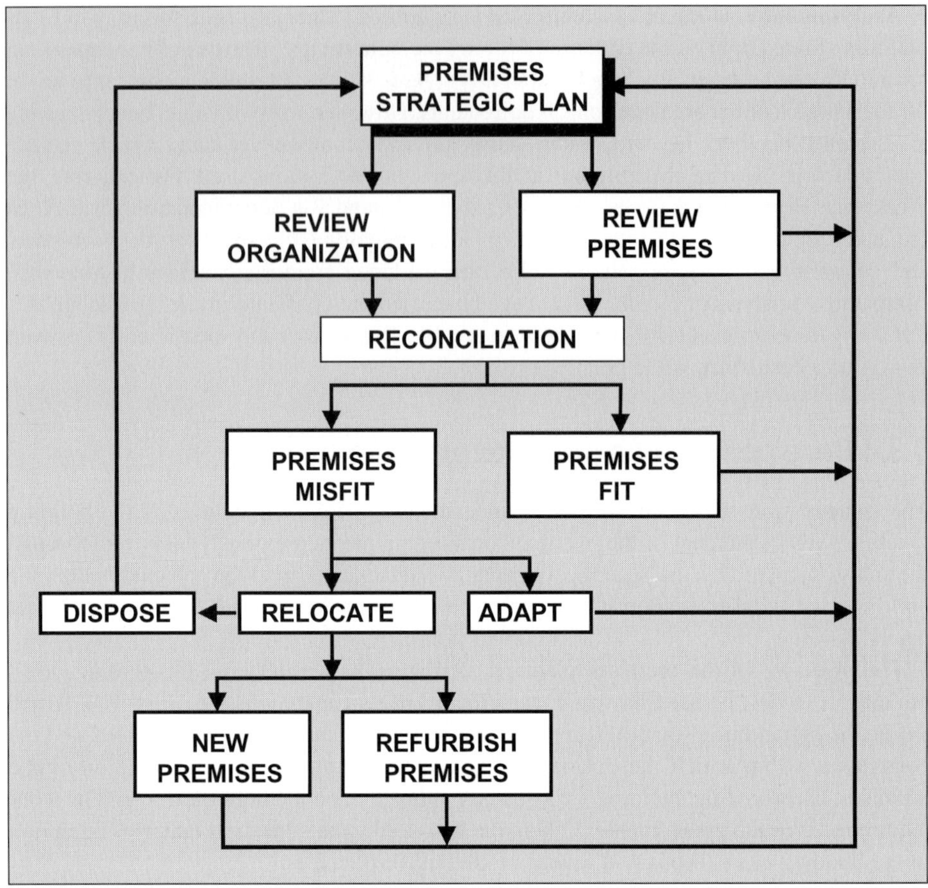

Fig. 3.7 A premises strategy review process.

- **Sub-division** – the scope that exists for sub-division of the workspace to enable parts of the building not required by the organization to be put at the disposal of other occupiers. This would embrace sub-distribution of utilities and services as well as the ability to create physically segregated workspace.

Once developed, the premises strategy should be set-out as a guide (hardcopy or in electronic form) for use by those engaged in planning and managing the use of the building.

3.3.5 Evaluation of hardware

From an analysis of the identified business needs, workplace standards and guidelines can be developed that will define the operational requirements of each workplace. The hardware evaluation process therefore starts with an appraisal of the existing products used by the organization in terms of their fitness for purpose to meet firstly current and then future needs.

The workplace assessments referred to earlier will provide information about the existing workplace settings and their effectiveness in meeting the needs of the users. These details, when set-out as specifications or performance standards for workplace components, will enable appropriate products to be selected.

With the vast array of products and components available on the market, it is essential that sufficient time is set aside to research adequately the products that may be suitable for the organization's needs. These products will include, amongst others, desking furniture, seating, storage systems, screens and partitioning systems. All of these can have a marked impact upon how efficiently the space in the building is used, through their influence on the 'footprints' of individual workplaces and space standards, as well as their contribution to the effectiveness with which the organization stores its information, materials, products etc.

It is important that the evaluation is undertaken from a standpoint of knowledge of the capabilities of the products and is based principally upon objective criteria. We recommend that, wherever possible, short-listed products should be tested in-use in the work environment, particularly workplace furniture, where experience has shown that live trials coupled with objective evaluation by the users can be especially revealing.

Once the existing products have been confirmed as being suitable, or new products selected, then *generic* workplace footprints can be converted into actual footprints based upon the intended hardware or furniture. At this stage it is worth checking the footprints as now proposed against those that were forecast at the analysis stage, since 'conversion' from the generic form, may necessitate some fine-tuning, which in turn may result in changes being necessary to the layout and area required.

3.3.6 Block and stack planning

Having determined the workspace requirements of the organization and the strategy for use of the buildings to be occupied, the next stage involves bringing them both together by mapping the space needs and work-group relationships of the organization onto the profile of the proposed building.

Block plans are prepared to indicate the 'footprint' that would be occupied by each work group arranged in relation to other groups, based on the affinities identified in the analysis. At this time the extent of enclosed and open workspace would be indicated on the building layout, as would the zoning of 'special areas' such as highly serviced spaces, process and manufacturing spaces, and common support spaces. *Stack planning* refers to the vertical arrangement of a series of such block plans that indicate the organizational structure, both vertically and horizontally, imposed upon a building profile of two or more floors.

Block and stack planning is best carried out in close consultation with customer groups and their nominated representatives, and is likely to require two, and some times more, iterations before a balanced solution is reached. It is good practice to have block and stack plans signed-off by the customer groups as being acceptable, before progressing to the detailed planning stage, thus avoiding abortive work on the more complex next stage. At this time, the approval from the work group (department, division or process team) is usually sought for the amount of space allocated, the mix of enclosed and open workspace and its location in relation to that of other groups. It is also recommended that where the planning exercise relates to a whole building or group of buildings, that overall approval of the proposed concept is sought from the senior management of the organization.

However, it is worth noting that having signed-off a block plan does not mean that the requirements of a customer group will not change, much to the apparent frustration of some facilities management teams we have encountered. As mentioned earlier, the organization never ceases to 'flex' in response to the ever-changing 'ebb and flow' of business. Consequently, there will be occasions when, after approval has been given to a plan, a work group may need to adjust their allocation of workspace. Hopefully, such instances can be kept to a minimum, not just to reduce the frustration levels of the facilities management team, but because the 'shelf-life' of the data collected will hopefully see the workplace planning through to its chosen horizon.

3.3.7 Detailed planning

Following approval of the block and stack plans, detailed plans can now be developed based on the concepts and principles indicated in the block, stack and footprint plans, and which include layouts for furniture, partitions, equipment, services, facilities and, in fact, everything that is required within the work environment.

Throughout all the stages, but particularly at the detailed planning stage, it is important that full visual communication of the proposals is presented to the intended occupiers of the premises. Many space planners, who have the benefit of training and experience are comfortable in reading plans and interpreting their intentions. However, this is not so of most workplace users who may have great difficulty in visualizing what the proposed workplace settings will look like. It is therefore essential that the detailed planning stage incorporates effective visual communication of the space planner's intentions. There are many means available to illustrate workplace layouts, which include three-dimensional drawings (manually prepared and CAD rendered images), as well as the use of mock-ups and visits to working installations in other organizations.

Whatever method is employed it is important to ensure, as far as it is possible to ensure the effectiveness of any visual communication, that the proposed workplace layouts and the ambience of the proposed environment meet the needs of internal customers and are acceptable to them. Again, as with the block plans, it is recommended that each work group, and possibly senior management, formally approve the proposed detailed workplace layouts, prior to progressing towards implementation.

3.3.8 Tendering and placing contracts

The process to be adopted for implementation will be dependent upon the scale and nature of the planning exercise as well as the resources of the organization. For some organizations this will entail using their in-house direct labour to carry out the necessary works, for others it will mean activating call-off contracts with previously selected preferred suppliers and contractors, but for most it will involve specifying, inviting and appraising tenders prior to placing contracts.

The processes and procedures to be followed for the tendering for goods and services are likely to be prescribed by the relevant purchasing and procurement function within each business, and hence are not subjects that it is worthwhile exploring here. However, there are some key points that should be considered by all organizations no matter how well they are served by in-house buying and contract services.

- **Competition** – in most organizations the tendency is to consider automatically that the supply of all products should be submitted to a competitive tendering process. However, there are often sound reasons for single sourcing products, such as continuity of specification, in-house production or technical compatibility, the absence of which (through the tendering process) may result in efficiencies and economies in-use not being achieved and, as such, act as a deterrent to effectiveness.

- **Specification** – tendering is best based upon technical specifications that are developed objectively and are known to meet the stated needs, which is easier to say than to achieve. Care should be taken to ensure that brevity is not achieved at the expense of clarity; however, the quality of a specification is not always in direct proportion to the voluminous size of the documents.

- **Cost** – be aware that 'cheapest is not always best', as cost-in-use is a much better basis for selection. This is particularly important in selecting hardware products that are purchased in the name of 'flexibility', where the cost of future demountability, relocation or reconfiguration should be considered at the tender stage, every bit as much as the initial purchase price.

- **Comparisons** – to enable selection to be based on price, as in the lowest tender, demands tightly drafted specifications so as to ensure that 'apples to apples' comparisons can be made. This is a frequent failing of many furniture tenders where generalized descriptions and generic 'footprints' are used to define the products and hence result in tenders being received where 'apples' are compared with 'oranges' and sometimes 'elephants'!

3.3.9 Implementation

Clearly, the method of implementation will be governed to a large extent by the nature of the work to be undertaken and the previously described tendering process. However, irrespective of the scale of the planning exercise, all workspace re-organizations should be undertaken to a plan. Typically, the principal components of the implementation plan should include the following.

- **Communications** – the methods to be used to keep all relevant parties informed, to the appropriate level, about the activities involved in implementing the proposed work environment and its progress.

- **Quality assurance** – procedures for the control of documentation, its distribution and revision.

- **Master programme** – setting-out the sequence, timescales and dependencies of each activity involved, which may include any fit-out works, decanting of workspace, interim moves, delivery schedules, and a functional audit.

- **Move plans** – details of key dates and times, roles and responsibilities, arrangements for rubbish disposal before, during and after the move, crate delivery, specialist equipment removal, labelling procedures, access and security arrangements.

- **Discontinuity** – how potential disruptions to business operations are to be minimized during the re-organization and its related works, the risks involved and the development of appropriate disaster recovery and contingency plans.

- **Variations** – the process and procedures to be followed by which variations to the workplace layout and its environment, for whatever reason, are to be presented, assessed and actioned.
- **Reporting** – the content, structure and frequency of reports from contractors and suppliers to the facilities manager, and similarly from the facilities manager to his or her line manager and the internal customers.
- **Handover** – the procedure to be followed for the handover of every element of the work environment from each individual contractor and supplier to the facilities manager, and then the handover of the 'composite' work environment to the internal customers.
- **Completion** – the recognition that practical completion is not the same as actual completion, and although handover and use of the work environment may be possible once practical completion has been achieved – a job is not complete until it is complete – i.e. actual completion is only achieved once all snagging and defect rectification has been carried out to the satisfaction of the facilities manager.

One final point about programmes borne out by experience. There is absolutely no benefit, no matter what managerial pressures are brought to bear, in setting unrealistically short timescales for implementation. Not only is it unlikely that the work can be achieved; but even where it can, it is likely to result in more defects and remedial actions than if it had been executed at a more sensible pace.

3.3.10 Functional audit

We are all familiar with financial audits, which are undertaken to assess whether or not the organization has managed its funds appropriately, that sums spent have been accounted for, and all is in compliance with statutory and corporate requirements. Very rarely do organizations ever carry out a functional audit, yet we believe the benefits that can be derived from such a procedure can be every bit as valuable.

To obtain maximum benefit from such an exercise, it is recommended that no sooner than 3 months, and no later than 9 months, after the new workplace layouts have been installed, a functional audit should be carried out, which has two components each with its part to play.

- **Post-Occupancy Evaluation (POE)** – the purpose being to ascertain detailed and objective feedback from the users of the workplaces on the effectiveness of the layouts, facilities and services in support of their activities. The findings should be reported with any shortcomings identified forming the basis of fine-tuning which may be required. Many organizations restrict their feedback to the POE only (see Fig. 3.8).

- **Process effectiveness** – adds to the POE since it is not solely concerned with the degree to which the environment and its services meet the needs of users, all of which is unquestionably important, but it also includes a review of the processes by which the planning was carried out and the environment created. It will include a review of timescales, communications, information flow, in fact the whole experience for the customer and supplier alike.

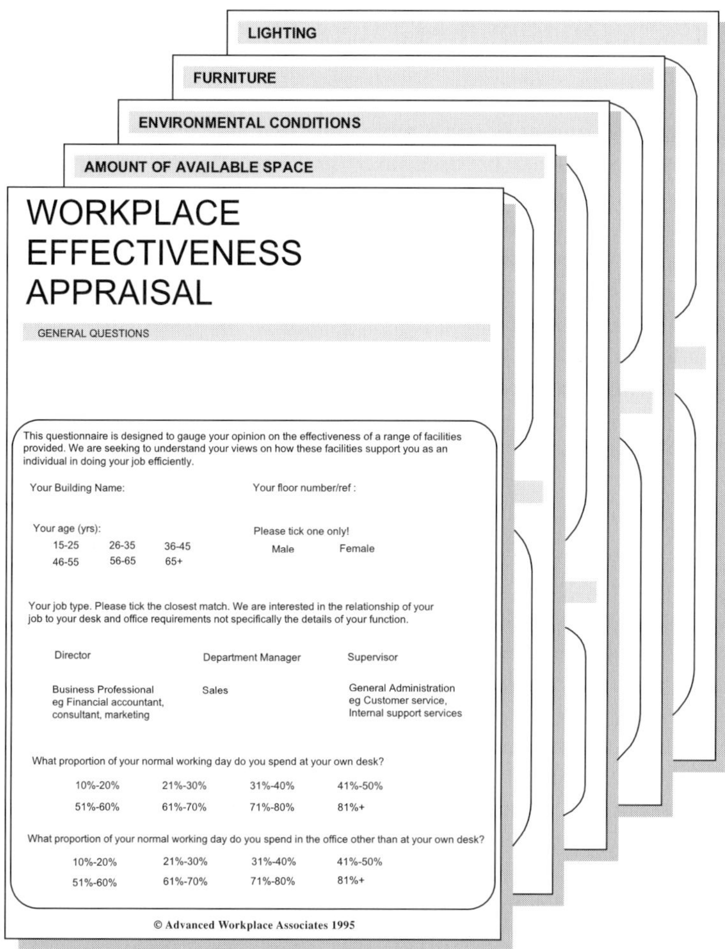

LIGHTING

FURNITURE

ENVIRONMENTAL CONDITIONS

AMOUNT OF AVAILABLE SPACE

WORKPLACE EFFECTIVENESS APPRAISAL

GENERAL QUESTIONS

This questionnaire is designed to gauge your opinion on the effectiveness of a range of facilities provided. We are seeking to understand your views on how these facilities support you as an individual in doing your job efficiently.

Your Building Name: Your floor number/ref :

Your age (yrs): 15-25 26-35 36-45 46-55 56-65 65+

Please tick one only! Male Female

Your job type. Please tick the closest match. We are interested in the relationship of your job to your desk and office requirements not specifically the details of your function.

Director | Department Manager | Supervisor

Business Professional eg Financial accountant, consultant, marketing | Sales | General Administration eg Customer service, Internal support services

What proportion of your normal working day do you spend at your own desk?

10%-20% 21%-30% 31%-40% 41%-50%
51%-60% 61%-70% 71%-80% 81%+

What proportion of your normal working day do you spend in the office other than at your own desk?

10%-20% 21%-30% 31%-40% 41%-50%
51%-60% 61%-70% 71%-80% 81%+

© Advanced Workplace Associates 1995

Fig. 3.8 A workplace effectiveness survey can be a vital part of the Post-Occupancy Review process.

We would argue that a functional audit is even more important than a financial one, since yet again, it is not the initial cost that is important to the company but the cost throughout its use of the facilities. Questions should be asked and answers found for issues such as 'Does the work environment that has been provided meet the original brief?' and 'Have the objectives of the accommodation plan been achieved?' 'Are the procedures adopted for space management effective?' and, for organizations that experience high levels of churn, 'Do the facilities provided for flexibility actually work in practice?'.

A cynic might say that the reason facilities managers do not carry out functional audits is based on fear of what the results might indicate. However, the main reason many of the organizations we encounter do not undertake audits is because they do not allocate sufficient, if any time, in the overall accommodation planning process to enable the review to be carried out and the findings analysed and reported. While the audit process should not prove to be too time consuming, certainly not in relation to its value, it does need to be consciously planned into the programme.

3.4 Roles, responsibilities and relationships

The process of planning and managing a business's workspace is likely to involve many people, and affect and impact on even more. The numbers involved will be a function of the size of the exercise to be undertaken and the way the business is organized. However, from our experience with many public and private sector organizations in the UK, Europe, North America, Australasia and the Pacific Rim, we have found a common pattern emerges. All of those that are effective feature the 'three Rs' of the workspace planning process – Roles, Responsibilities and Relationships – which are fundamental to the success of any accommodation planning process. Below we set out the key roles and responsibilities that need to be effected to bring about successful planning, and which we have found feature in every organization that demonstrates well-managed workspace.

3.4.1 Leader

The leader should be the champion of the planning exercise to be undertaken. Often a business leader or manager, ideally he or she should be the most senior executive who will be involved in the re-organization, i.e. whose work environment is to be changed. However, as such personnel need to be 'visible' and 'audible' in the cause of the re-organization, then they must also be available, which is not always easy to guarantee with senior personnel, so personal commitment is also a vital attribute of a leader.

The leader will be required to attend occasional project meetings and may also, where the nature of the exercise so dictates, chair a 'steering group' who will provide overall corporate guidance for the project. On large-scale projects, such as the development of new premises or major relocations, the leader is likely to be responsible to the Board or Executive of the organization for the overall success of the exercise.

3.4.2 Facilities manager

The facilities manager will usually be the person who will be assigned overall responsibility for the provision of the workspace, its planning and implementation. On many exercises the facilities manager will be the provider of much of the information about the buildings occupied by the organization and their infrastructure, and will be the custodian of the information gathered as a consequence of the planning exercise.

Most often the facilities manager will chair the project team charged with delivering the planned workspace to the organization. He/she will be responsible for the day-to-day activities of the planning exercise and will act as the link between the internal customers (the workspace users) and those who will carry out the accommodation planning activities. However, care must always be taken to ensure that the allocation of responsibility to the facilities manager for significant re-organizations is not achieved to the detriment of his or her 'normal' activities. Where this role is carried out by the facilities manager, then the role and responsibilities described below for the project manager would also apply to the facilities manager.

3.4.3 Project manager

On large-scale workspace planning exercises or where the facilities manager does not have adequate resources available, then it is likely that a project manager will need to be appointed to carry out the tasks that the facilities manager would otherwise have to address. The project manager, who could be an in-house appointment or an external consultant recruited for the exercise, will have responsibility for the day-to-day activities of the accommodation planning exercise, and will be accountable for delivering the work environment within the agreed targets of time, cost, quality and safety.

He or she will lead the project team comprising professionals (in-house and external consultants and contractors) who will tackle the various planning stages appropriate to the nature and scale of the exercise. The project manager will have overall control of the budget and the programme, and will set the timescales and sequencing for many of the activities. On very large or complex re-organizations, the project manager may be supported by a move manager, who will act as a 'sub-project manager' tasked with the planning and implementation of workplace moves, relocations and any interim decanting that may be required. On most accommodation planning exercises these activities will remain within the remit of the project manager.

3.4.4 Business planner

The business planner's involvement in the accommodation planning process will generally be in relation to the strategic planning element, where he or she is likely to be the primary source of information about the future direction of the organization and its business. In some organizations this will be the province of senior management, e.g. the CEO or a director, in others – particularly large corporations – then the business planning activities are usually carried out by a dedicated team of people.

The business planner's role will be to provide the business scenarios and supporting information, referred to earlier in this chapter. In organizations where no formal business scenario modelling takes place, then the business planner will assist in the development of possible scenarios by providing the relevant information. The process of developing possible accommodation planning scenarios is likely to involve a high level of interaction between the business planner, senior management of the organization (probably the corporate client) and the facilities manager, in order that the nuances of each scenario are fully tested.

Because the information that the business planner makes available is obtained from his or her 'crystal ball' it does not mean that it is vague, general or imprecise, on the contrary it needs to be very specific and detailed, and the business planner will be responsible for ensuring that it is as reliable as it can be.

3.4.5 Researcher

The researcher is a vitally important member of the accommodation planning team. On all but the smallest of exercises the data gathering will most probably involve more than one person. Researchers will be responsible to the project manager for the successful collection, collation and analysis of the required data upon which the accommodation planning will be based, often working to very tight timescales wedged between the requirements of the project programme and the 'windows' in peoples' diaries. They need to be effective communicators and that includes listening, as they need to be able to hear what is being said and also to articulate what is *not* being said, so that comprehensive data are collected. They need to be skilled in all of the data gathering techniques, e.g. interviewing, workshop facilitation, surveys and observation. Frequently, they will be required to conduct research into specific work-related topics, operations at other organizations (case studies) and benchmarking comparisons.

On many planning projects the research functions will be carried out by the space planner in addition to their role as described below.

3.4.6 Space Planner

Second only to the customer, the space planner is possibly the most key member of the accommodation planning team charged with converting the organizational and building data and its analysis into proposals for a workplace environment. The space planner will be responsible to the facilities manager or project manager depending on the project requirements, for producing workplace layouts that meet the needs of the customers – by exploiting to the full the capabilities of the premises and infrastructure available to the organization – within the set budgetary constraints.

Although the tangible evidence of his or her work, and the means by which his or her efforts will be judged will be the work environment that is created, most of the activities of the space planner will be centred around the production of workplace layout drawings and designs. However, in addition to technical skills, the space planner needs to possess excellent communication skills and be adept in both oral and visual communication techniques. He or she needs to be able to explain his or her proposals to people for whom workplace planning is a new experience, in a non-threatening and inclusive way. The space planner needs to be able to articulate the benefits and the risks in adopting the proposals, which can be extremely challenging when what is being proposed involves the customer in significant change, such as moving from enclosed to open workplaces, or from dedicated to non-dedicated workplace provision.

3.4.7 IT provider

The role of the IT provider is that of both technical specialist and visionary. In his or her capacity as technical specialist s/he will be responsible to the project manager for the specification, design, planning and implementation of the voice and data systems required to support the organization's work environments. The IT provider will work closely with the space planner in developing integrated workplace solutions to meet the needs of the customers.

In his or her visionary role, the IT provider will assist both the business planner and the facilities manager to explore the opportunities that exist to 'reshape' the organization by applying technology-based solutions and tools. These tools could have a significant bearing upon the numbers of people required by the business, their activities, their location and their working practices, and hence could have a major impact upon the accommodation planning scenarios that the organization employs.

Both roles, i.e. technical specialist and visionary, could be discharged by in-house or external parties or a combination of both, depending upon the organization's resources. However, the visionary role is more likely to involve strategic business input and hence, if not provided from in-house resources, needs to be 'tempered' by the business planner.

3.4.8 Customers

Internal customers, a term we much prefer to the widely used 'users' label, are collectively the people who occupy the building and use the work environments contained within them. Customers will therefore be departments, working groups and individuals, as well as the organization as a whole, i.e. the corporate client as described below.

Customers will be providers of valuable information about their work and work process needs, and can also provide crucial insights into how effective the existing workplaces are in supporting current business operations. As in any other market, the community who occupy the organization's premises can be segmented into different groups based upon their use of the buildings and their workplace needs. Consequently, when gathering data about the requirements of customers it is necessary for the researcher to be alert to this 'segmentation' as what may appear superficially to be customers with the same requirements may, on closer investigation, turn out to be people who have quite different needs.

The quality, in terms of functionality, of the work environment to be created will be highly dependent upon the quality of information provided by the relevant customers. Therefore, customers must make sufficient time available to participate actively and constructively in the planning process – who knows they might even enjoy it, it has been known! Their participation will encompass the provision of data, the sharing of experiences and, on occasions, even carrying out their own research, all with the objective of providing the planning team with the most comprehensive and reliable information that is available.

In almost every planning exercise, customers will channel their requirements through a nominated representative, whose responsibility it will be to act as the link between the department or working group and the accommodation planning team. The customer representative will therefore be responsible for providing promptly, the relevant data required by the researcher, although he or she may not be actively involved in collecting the data him or herself. Similarly when the space planner presents proposals for the workplace environment, it will be discussed with the customer representative who will then be responsible for disseminating the information to the work group that s/he represents. From this it should be clear that the customer representative needs to be knowledgeable about the work activities of those s/he represents and have the authority to speak on their behalf.

3.4.9 Corporate client

The corporate client may be the leader or the facilities manager and might also be a customer, and where this is the case then the corporate client would fulfil those roles as defined above. However, the corporate client also has another part to play in the accommodation planning process, which does not always feature in the aforementioned three roles, and that is as custodian or guardian of all things corporate. For an organization that occupies a single building the corporate client would almost certainly be involved in the planning process, through discharging one or more of the roles described above.

The corporate client could be one individual or a group, such as a committee, who may not be actively involved in the specific building or workplaces that are being planned, but who have a custodial interest in their provision and possibly their use. They may be the Landlord, literally, or act in that capacity, they may be the final arbiter on image and market positioning, they may be the financial authority for the business or they may be the unifying agent in a multi-site, multi-building organization. Whichever of these, or other capacities they act in, the project manager needs to ensure that the corporate client is consulted as and when required, and kept informed throughout the entire accommodation planning process.

3.4.10 Implementer

The role of the implementer is to create physically the required work environment within the chosen premises, to the standards and arrangements as defined by the space planner. This may involve in-house direct labour in moving existing work settings, or suppliers installing new workplace furniture, through to contractors undertaking fit-outs and the refurbishment of buildings, and even the construction of new premises. Frequently, the implementer will comprise several organizations whose combined specialist skills and collective efforts are brought together to produce the desired environment.

The implementers will be responsible to the project manager for the delivery of the required work environment within the set parameters of quality, safety, time and cost. They will need to work closely with the space planner to ensure correct interpretation of the workplace layouts and designs.

Because no implementer has the capabilities to produce the complete work environment – building, IT systems, furniture, plant and machinery, etc – then the challenge for the space planner, and ultimately for the project manager, is to coordinate the input, products, services and activities of all the separate implementers with a view to delivering a composite and cohesive workplace to the customer.

3.4.11 Property agent

The role of the property agent is to find suitable buildings that match the needs of the organization, to arrange for their acquisition (lease or purchase), to advise on the terms of occupation of the premises and to dispose effectively of redundant premises.

When the facilities manager is considering how best to address the requirements of different scenarios in the accommodation plan, then the property agent can assist in conducting research into the premises options available and the possible courses of action that could be adopted.

The property agent will be responsible to the facilities manager for the identification of potentially suitable buildings. The brief for the property search will have been fashioned from the statement of requirements identified through the analysis of the business's accommodation needs and will be used to short-list potential premises for further, more detailed investigation by the facilities manager and the space planner.

3.4.12 Other roles

The roles described above relate to those most frequently involved in the selection and creation of a business's workspace. That is not to say that there are no other parties who could and should participate in the process; this will be dependent upon the specific nature of the accommodation planning exercise being undertaken. A short list follows of other possible participants and the roles they can play.

- **Lawyer** – the adviser on all things legal, which will include contracts, leases, property titles and the procedures to follow in acquisitions and disposals.
- **Architect** – the designer of a new building or the fit-out or refurbishment of an existing building, some may also provide space planning services.
- **Interior designer** – the designer of the interior environment of the building, some may also provide space planning services.
- **Work-process designer** – can assist in the mapping of work activities and the design of work processes.
- **Ergonomist** – specialist in ergonomic design of workplaces and the 'hardware' contained within them.
- **Engineers** – a range of specialists who have skills in building structures, mechanical services, electrical services, lighting, acoustics or production engineering.

The individual roles of the people involved not only have an important part to play in the delivery of elements of the accommodation planning process, but by the manner in which they are brought together and the way their respective inputs are managed, they can also have a major bearing on the overall success of the exercise.

3.5 Managing the accommodation planning process

All too frequently, in response to pressures of time, those responsible for planning workspace demonstrate well the old adage that 'what is done at haste is repented at leisure'. Their dash to get on with planning the work environment is achieved at the expense of proper data gathering and analysis of users' requirements. This is often exemplified by the push to buy furniture products (possibly because of quoted long delivery times or imminent price increases) before understanding the needs of the users and the capabilities of the products – this can result in them acquiring tools that are inappropriate for the jobs being asked of them. By observing and investigating how people really work, then new and unexpectedly beneficial results can be achieved.

We are all limited by our own experience and, in the context of work environments, it is therefore incumbent on all of those responsible for planning workspace to develop their own understanding of the 'art of the possible' and the 'risks of the improbable', and to share these experiences with customers. This is not done with a view to squeezing them into inappropriate workspace solutions slavishly copied from other locations, but with a view to enabling customers to consider the options that may exist for doing things differently. In such situations the roles of the facilities manager and space planner should be to facilitate new thinking by demonstrating options and alternatives of which the customer is unlikely to be aware.

3.5.1 Workplace pilot

When new and innovative approaches to the workplace are being contemplated, then we believe these should first be tested by way of a workplace pilot programme. This involves setting-up part of the work premises in the proposed configurations in which selected personnel will work, so that in-use feedback can be obtained from customers, through the use of the appropriate criteria and tools to evaluate the effectiveness of the pilot layout. The criteria used for evaluation should include measures relating to the ease of use and comfort of the workplace as well as productivity assessments.

While for those on the project team the pilot is the focus of their activities – a 'live' workplace laboratory – for others not involved in the development process it could be viewed as a nuisance or perceived as a threat. Therefore, the implementation and monitoring of the pilot must be conducted in a very controlled manner so as to avoid disrupting the ongoing operations of the business, and alienating those for whom the pilot will be a 'distraction'.

Experience has shown that such pilot programmes should ideally run for 6 months, but not less than 3 months, to ensure that full value is obtained. The evaluation of the pilot should be an ongoing process commencing from its implementation, and based upon the previously determined objective performance measures. However, most of the evaluation is likely to be conducted during the last 4 weeks of the pilot, at which time the data gathered by questionnaire, interview and observation should be collated and analysed against the success criteria previously set and incorporating any other issues identified along the way. The findings of the pilot should then be set out in a report to business and facilities managers, detailing the recommendations and future actions, if any, that should be followed.

3.5.2 Communicating the accommodation strategy

Effective communication of the proposed accommodation strategy is of paramount importance for its success. Each organization should effect communication using its existing channels wherever possible. It should be the aim of those responsible for the accommodation planning process, that representatives of all work groups be kept informed of the status of the strategy throughout its development. Of course, the communication process begins before the strategy has been developed. It commences as soon as the business decides to undertake the preparation of a strategy, and continues throughout the data gathering stages, as those involved in the process interact with people in the various parts of the business.

Of critical importance for success, it is vital for all of those involved to recognize that communication is a two-way process. Far too often, those involved in preparing and communicating a strategy, flushed with the excitement of their endeavours, embark upon a monologue, forgetting to encourage participation which will provide them with an opportunity to listen to contributions from others.

Our experience has shown that the introduction of changes to workplace strategies is best conducted via workshops where affected personnel are taken through the key components of the work environment and its operation in an interactive process. Not only does this approach lead to a better introduction of the components of the strategy, but it also fosters a greater prospect of receiving 'buy-in' from key customers. Such workshops, which would normally be of 2–3 hours duration, should be attended by all customer representatives and other parties involved in the planning process. At these sessions, the accommodation strategy and its implementation would be described and explained by a facilitator who would lead the representatives through the process of any proposed pilot programme, to encourage them to use the pilot as an opportunity to view the new work environment in operation, and also answer questions that attendees may have.

3.6 The use of computer tools

The amount of data required, for even the most modest of accommodation planning exercises, can be extensive. The use of appropriate computer tools can therefore be of great assistance, if not a vital necessity.

The role of the facilities manager is correspondingly complex, requiring control over a wide range of diverse activities, which are necessary to support the organization in achieving its business objectives. A key aspect of this role is the ability to access, manipulate, store and report information that is fundamental to successful management. Therefore, to stand a chance of meeting the organization's requirements, those responsible for facilities management need to be skilled professionals equipped with appropriate tools – this is where Computer Aided Facilities Management (CAFM) tools can play an important role.

To turn a facilities management group into a 'learning organization', to coin a term from Peter Senge's (1990) book *The Fifth Discipline*, it must pay attention to the ways in which it encourages learning and growing. As defined in the *Fifth Discipline Fieldbook*, 'learning in organizations means the continuous testing of experience, and the transformation of that experience into knowledge – accessible to the whole organization, and relevant to its purpose.'

The computer services industry has responded to these demands by producing a range of products and systems which are claimed will help the hard-pressed facilities manager. But how do you select a system that is appropriate for your needs? What products are available and how would they integrate with existing manual and computerized systems? How would you assess if the tools will provide you with value for money? How should they be cost justified?

The route to finding the answers to these and many other related questions, is through identifying where facilities are placed within the organization's priorities and the seriousness with which they wish to address the use and protection of facilities as assets. From this can be determined the need for different types and quantities of information to suit strategic and tactical needs as appropriate, as not all CAFM tools are structured to meet the needs of either, and very few to meet the needs of both.

3.6.1 CAD, CAFM and other tools

Although frequently bracketed together under the banner of CAFM tools, it is important to realize that the different types of computer tools available are intended for specific purposes, each with their own attributes. They include: real estate and property management, surveying and recording, building management systems, maintenance, space and organizational planning, inventory and tracking, financial control, contract management and many more.

In determining the structure of the database required to meet the organization's needs it is important to distinguish between quality of data (relevance and reliability) and quantity of data. The nature of the business and the type of assets will, to a large extent, determine the core data set upon which operational processes are dependent. But beware. All too often facilities management teams, flushed with the enthusiasm of obtaining new tools, embark upon a programme of inputting enormous volumes of data, in the mistaken believe that 'everything' needs to be 'on the system'. The results are reasonably predictable. Either resources (if not enthusiasm) will run out, leading to incomplete and hence irrelevant databases or, where permitted to run their full course, will result in 'analysis paralysis' where the facilities management team are submerged in data with not a drop of information in sight.

For many organizations, the perceived benefits of acquiring a CAFM system are often not fully realized, usually as a consequence of inappropriate or insufficient training of the intended users. It has often been said that the pace of technological change is fast and unstoppable. However, in practice, the pace of change is dictated by how fast human attitudes and behaviour changes – giving rise for the need for education and training. Implementing CAFM tools is no exception; where the need to train is a crucial foundation upon which the long term success of the tools, and of the facilities management operation itself, is based.

CAFM product selection must of course be based upon meeting current business needs. However, of equal significance in system selection, is the product's ability to provide a path for conversion or migration of the tool to meet changing future needs. In addition to coping with changes in response to the inevitable flexing of the core business, it is important to realize that changes in the way the tools will be used are likely to occur as operators become increasingly skilled, and the organization becomes more demanding in its expectations of the management of its facilities.

It is a vital prerequisite of implementing any CAFM system, that it is introduced into a *managed* environment. Full benefit is rarely achieved when the use of the tools is restricted merely to automating existing paper-based processes. The drive to computerization also offers an excellent opportunity for re-engineering the existing facilities management processes, the focus of which should be on how the new technology can lead to new, more economical methods of achieving service provision. At the very least it will be necessary to assign responsibility for the CAFM system to an individual. It could also mean realigning the roles and responsibilities of the facilities management team, as the tools may change the way the team operates. There will be a need to define responsibilities relating to the use of the system, the data capture, entry and updating processes, and access to data and reports generated by the tools. Through careful identification of business needs, objective product evaluation and a disciplined approach to its implementation, the use of CAFM tools can make a meaningful contribution to the effectiveness of the management of an organization's accommodation and facilities, and in so doing add value to its core business operations. In considering information support to workspace and churn management, it is crucial to regard data handling and data management as integral components of the overall asset management process, rather than as administrative overheads. For many, this will require a fundamental shift in mindset from regarding data as a problem, to considering it as a resource that can be turned into a useful asset of the organization. It is for this very reason that justification can be made for investing in computerized management information systems. Some relevant pointers in assessing data and information needs include the following.

- Establish key workspace management processes and core data elements.
- Establish key decisions at strategic and operational levels, and their output format.
- Prioritize key deliverables within the implementation programme.

There are very many types of computerized tools that can be deployed by the facilities manager, and which tend to get bracketed together under the banner of CAFM. However, they are wide-ranging in their capabilities and in their application. Of the many product types currently available, the following tend to be the main generic categories.

- Asset inventory and tracking.
- Maintenance – (PPM).
- Survey and recording.
- Building Management Systems (BMS).
- Space planning – organizational planning and layout planning.
- Facilities reservation – meeting rooms, equipment and hospitality services.
- Financial control – budget and expenditure control, and charging.
- Security management.
- Real estate and property management.
- Contract management.
- Project management.

Many of the products are derived from a computer aided design (CAD) package that is used as the core around which the facilities data are developed.

Computer aided design is an unfortunate name for something that can be used both as a management as well as a design tool. As with all other computerized techniques, CAD affords the facility to handle large amounts of data very easily, with accuracy and speed of response thrown in. When utilized to its full extent it can have the added benefit of an enhanced communication tool through using three-dimensional modelling techniques, which can be of immense benefit for both strategic and tactical planning.

For all but the smallest and simplest of office buildings, manual techniques for space planning are no longer appropriate. Nonetheless, there are still organizations that have yet to realize the potential benefits from servicing their needs through computerized techniques. As the cost entry thresholds have plummeted through the 1990s, the payback period has shortened too, in some cases, to as little as 12 months. As a facility providing readily updateable drawing records, it can be cost justified for most organizations who have a churn rate in excess of 20%.

Using CAD systems can permit the development of a graphical database of premises and their contents. In many organizations, this may give rise for the first time to fully documented records of premises layouts, services distribution routes, furniture layouts etc. Even if they pre-exist, a CAD system will make them more manageable, accessible and more easily updated. Through the 'layering' capabilities of CAD systems, different layers can be assembled for different elements such as HVAC systems, IT services, lighting, partitions, furniture and equipment, etc. These individual layers can then be combined and printed in different combinations that have the potential to make it a very valuable facilities management tool. Further, where such graphical databases are interlinked with non-graphical databases, then data can be assigned to every graphic symbol producing an even more powerful facilities management tool. By attaching to the drawing of a desk, the details of the manufacturer, the product range, the date of purchase, the supplier, the price etc, it is easy to see how extensive databases can be established and, by simple manipulation, schedules and management reports can be generated from 'drawings' held on the system.

Some CAFM tools take the elements of CAD further (see Fig. 3.9). A good CAFM system will possess open software architecture, to permit the import and export of data files from other software, some of which may pre-exist in the organization. It is also likely to be fully interactive and integrate the data of personnel records, lease management and other cost-based management information, with the graphics associated with the CAD system. While for many organizations CAD will be of benefit, non-CAD CAFM systems are available for those who do not need the graphics management aspect. With these systems and techniques, the facilities manager and his or her team, have powerful tools which, in the right hands, and following a well developed accommodation plan, can make

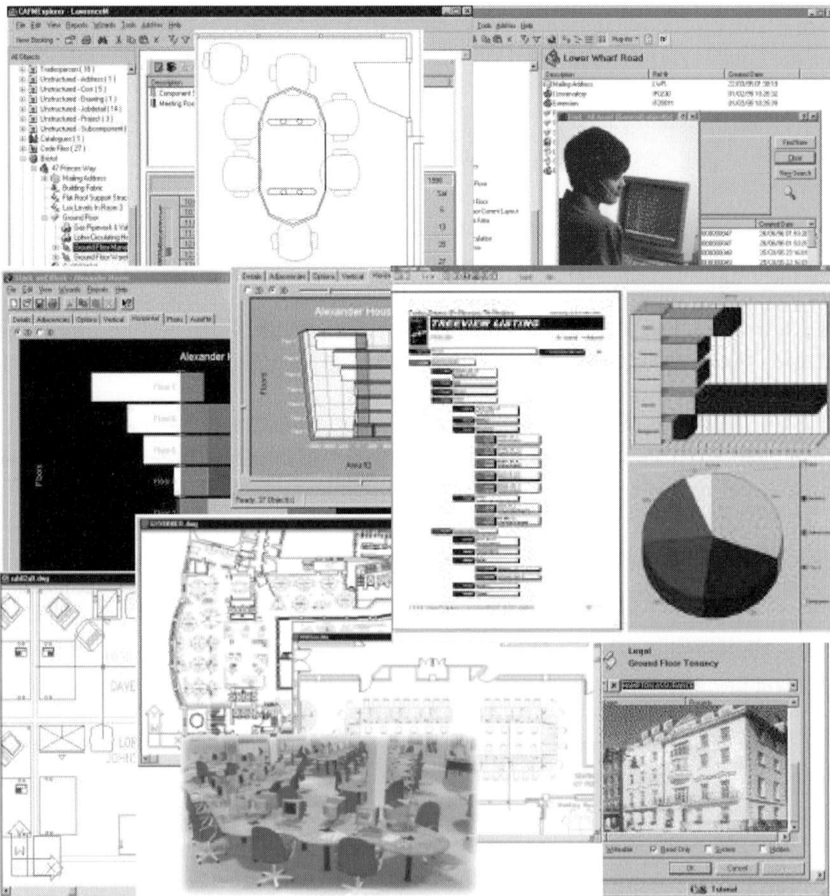

Fig. 3.9 CAFM tools can address a range of needs from strategic planning and scenario modelling to space planning, tracking and asset inventories (source: AutoFM by Decision Graphics).

possible the apparently impossible and certainly permit more reliable forecasting to be achieved.

With most CAFM systems, the data builds into a 'model' of the organization. This model can then be tested and examined under different conditions. By varying the data, 'what if' scenario modelling can be achieved, allowing the implications of possible scenarios to be considered and planning revised accordingly. For example, many systems will automatically re-stack the building model when a department has been changed or relocated to show how it will impact on others. The speed and flexibility of such powerful databases means that the facilities manager can consider many accommodation options before deciding on the appropriate course of action to adopt.

The big need for many users is to be able to access data quickly and to be able to convert that data into management information when appropriate. Often the data must be shared between applications or sites. Developments in hardware, application software, and cabling are enabling all of this to happen. Increasingly, more 'user friendly' tools and software applications are being developed with many of the following features.

- Microsoft Windows versions of packages.
- More interfaces between applications, with better connectivity.
- Features to make user customization more practical.
- Reports that are easier to generate.

Increasingly important in the assessment of products, particularly if you are going for a modular application approach involving different manufacturers' software, is ensuring they all 'talk the same language'. This means not only that they are all software compatible, i.e. operating on a MS Windows environment, or capable of importing drawings in dxf file format, but also that they have adopted the BIFM or IFMA protocols or, if not, that they are capable of easy adaptation to enable these protocols to be used.

3.6.2 Management of a CAFM System

The person assigned the responsibility for the system should have his or her role defined to include the following.

- Responsibility for putting forward the case for CAFM in meetings of the facilities management team, and for considering how use of the system can help the team in any new initiatives that may arise.
- Preparing a strategy for the use of CAFM, which should define the boundaries, look at the present situation and envisage future developments, particularly the almost inevitable linking with other computerized tools.
- Preparing an implementation plan not just for the new system but also for any upgrade or additions to it in the future.
- Monitoring of the objectives set prior to acquiring the CAFM system: in the early days this will have two dimensions – (a) as progress against expected goals compared with traditional methods, and (b) as hoped for progress set by expectations of the system. As time goes on, and experience shows relative strengths and weaknesses, one will replace the other and eventually the first will disappear altogether.

- Responsibility for setting a budget for the running of the system and any support services that may be required, including training, data capture, maintenance and additions to the system.

The purchase of a CAD or CAFM system is only a part (and in some cases a small part) of the total cost of providing such a system. Most organizations grossly underestimate the cost of loading up the database so that their system can be brought into effective use. In many instances, the best way for text and graphics to be loaded into systems is by manual entry. Of course, it may be possible, where the information is already held electronically, to import files directly from a software package into the CAD/CAFM systems if their open architecture permits.

Today's business world of constant change and ever increasing complexity, demands that competitiveness be achieved in all aspects of the organization's performance. The management of the organization's accommodation and facilities is no exception – the need to be both effective and efficient is paramount. By way of illustrating some of the issues raised in this chapter, we outline the following brief case notes which demonstrate quiet different approaches that some organizations have adopted as part of their accommodation planning processes.

Case Note 3a

Oil company, Aberdeen, Scotland. At the start of an assignment for the planning of a single new operations centre to accommodate around 900 staff previously located in four buildings, the General Manager responsible for the operations challenged why, in our attempt to develop a brief of requirements, 'is it necessary for you to speak to the staff?'. His assertion was that we could see the company's current workspace, so 'just give us more of the same but newer and in one location'. This was the expression of a person who sees the data gathering process as irrelevant, misleading and a distraction to his people and his business. Our response to him, had we been quick enough in thought, might well have been 'why don't you drill for oil in my garden, and never mind the exploration?'. Of course both would be equally wrong. The value of the data gathering process is that it seeks to establish how people work, what their experiences have been in their current environment, and how it is considered that their working practices may change in the future – none of which can be achieved without 'exploration' and 'speaking to people'.

Case Note 3b

International conglomerate, London. In developing their brief for the work environment of their new world headquarters, this major business conducted a survey of all the 500 staff to be relocated from three buildings, to ascertain their real needs and preferences for their workplaces and support facilities. The 60 item questionnaire was issued to all staff simultaneously, giving them 3 weeks to complete and return it for analysis. A 60% response was received, the results of which formed an integral part of the brief for the development of the new building. A comprehensive summary of the survey's findings was given to every member of staff as a feature of good communications.

Case Note 3c

Business machines company, Southern England. In considering how to improve the performance of the staff in their European Headquarters through better communication and a more effective work environment, the CEO saw benefits to his own working and that of his senior management team, by the introduction of new flexible styles of working and non-dedicated workplace provision. He fully embraced the concept and led by example not only through adopting the approach for the senior management team, but by being the first group in the organization to embark on such working practices, leading from the front – a championing leader.

Case Note 3d

Insurance company, London. When considering the strategy for the use of their existing London premises senior managers could see the benefits to the organization's cost base and productivity levels by the implementation of flexible working practices for all staff, but chose to exclude themselves by retaining their grandiose 'baroque offices'. This they attempted to justify, by claiming that their actions had nothing to do with status, but was required because the environments were 'impressive to their customers'. Impressive they undoubtedly are, but effective and representative of a progressive business? By their actions, we believe they demonstrated that they were out of touch with their staff and customers and did not really understand the true benefits, to themselves and the business, of the improvements to collaborative working and individual effectiveness which they so freely espoused.

Case Note 3e

Pharmaceutical company, Southern England. The research and development department of a major pharmaceutical company was being re-located. The scientists who develop the drugs that deliver vast profits to such businesses required very high specification workplaces, which in the past had been provided by the creation of small, enclosed 'cells'. The intentions of the corporate client was that these should be repeated in the new location. However, as a result of discussing with the scientists how they worked and what led them to avenues of research, it became apparent that much duplication of effort often occurred because people would be carrying out similar lines of work oblivious to the efforts of others. Also, while not in itself leading directly to the breakthrough, it was considered that unstructured discussion amongst colleagues, in a relaxed and informal way, frequently led to spontaneous lines of thought which took research along paths that may well not have been explored, or at least not for some prolonged period. In other words, the opportunity for informal interaction contributed to business success, which was not the sole domain of more conventional work settings. So environments were created and procedures developed to encourage social interaction amongst colleagues, such as promoting 'coffee areas' in preference to drinking coffee at the workplace, and informal social spaces with soft seating spaces where staff could congregate and, in so doing, share their thoughts and experiences.

3.7 Conclusion to the planning process

The accommodation planning process described in this chapter is of necessity detailed and comprehensive, and may appear daunting. It is not envisaged that it will be necessary to engage in every element of the process for every planning exercise that readers may need to carry out. Rather, it is a process that should be followed to establish an accommodation strategy and, thereafter, its components used as required to suit the needs of specific planning exercises.

Once the knowledge base of the organization and its facilities management team has been established by following the principles and procedures described, then that base of information can be 'topped-up' by implementing a review cycle to capture new and/or updated information at both strategic and tactical levels. The timing of the cycle should, as previously stated, be linked to the business planning process followed by the organization thereby ensuring that the most accurate assessments of business needs are available and that these needs and their means of delivery are synchronized.

In the two chapters that follow we investigate how to set about balancing the workplace needs of the organization with the workspace that is available, i.e. reconciling supply and demand. First, in the next chapter, we consider the way to assess the needs of the organization – the demand side of the workspace equation.

References

Barrett, P. (ed) (1995) *Facilities Management – Towards Best Practice* (Oxford: Blackwell Science), 77–121.

Becker, F. and Steele, F. (1995) *Workplace by Design: Mapping the High-Performance Workscape* (San Francisco: Jossey-Bass) pp. 103, 109 and 123.

Edwards, B. (1998) *Green Buildings Pay* (London: Routledge).

O'Reilly, J.J.N. (1987) *Better Briefings Mean Better Buildings* (Watford: BRE).

Senge, P. (1990) *The Fifth Discipline: The Art and Practice of the Learning Organization* (New York: Doubleday).

4

Assessing demand – the organization's needs

As described in Chapter 3, best practice requires organizations to put in place a process through which the facilities manager and those responsible for workplace planning and provision, can identify the needs of the organization and assess how these will change through time. The aim is to provide premises and facilities that match the identified needs of the business, that is, reconcile the requirements of supply and demand.

The planning of all workspace should be a response to specific business needs and drivers, what we call here the *demand*. But how are the needs of the business expressed and by whom? Is it reasonable to expect business managers to articulate their needs in terms of workspace and infrastructure? Our contention is that a key, if not *the key*, aspect of the role of the facilities manager is the interpretation of business data into a set of requirements for workspace and its infrastructure. In this role the facilities manager acts as an interpreter between the business and its suppliers, translating the 'business speak' of managers into facilities language, to enable possible solutions to be developed, and then vice versa, to present workspace proposals for consideration by the business.

The data gathering process should therefore be structured to capture information about the current and the future needs of the business, and the people within it. In this chapter we seek to consider the issues that can arise when determining the needs of the organization and how they are to be gathered, what data are required and how analysis of these data can lead to the development of workplace solutions that truly support the business and the people within it. We will look at the types of information required, work and work processes, the geography of the workplace, workspace allocation and standards, and the process of forecasting future needs.

4.1 Types of information required

As we saw previously, the level of detailed information required will vary depending upon the stage of the planning process, i.e. strategic or tactical. Also, in certain circumstances, some processes and techniques are more appropriate to use than others to gather the data. It would, however, be unrealistic to expect business managers, unless they are unusually skilled in this area, to be able to provide the information that the facilities manager can use directly to plan the organization's workspace requirements. Therefore, however it is gathered, the information required by the facilities manager will pass through a process of conversion from business information into workspace data (see Fig. 4.1).

Typically, the types of information the facilities manager will require, to enable planning of the workspace to be carried out, will include the following.

4.1.1 People

- The types and numbers of people who will be using the building and its facilities, which in addition to personnel based at the building would also include other parties who are visitors for short or long durations.
- The numbers of people to be engaged in each of the work processes of the business.
- The anticipated occupancy of the premises in terms of numbers of people, timing and duration.

4.1.2 Workspace

- The amount of workspace that is currently required by the business and will be required in the future, expressed for the organization as a whole, and for each business unit individually. It is important to collect these data by the smallest organizational group, i.e. business unit or work group, to enable variations in need from one group to another to be assessed in terms of their relative impact upon each other and upon the total requirements of the organization; an increase in demand by one work group may be able to be addressed by the reduction in demand of another, without any net effect on the workspace needs of the business as a whole.
- The types of workspace required and its attributes, such as narrow plan, deep plan, column free and the amount of contiguous space.

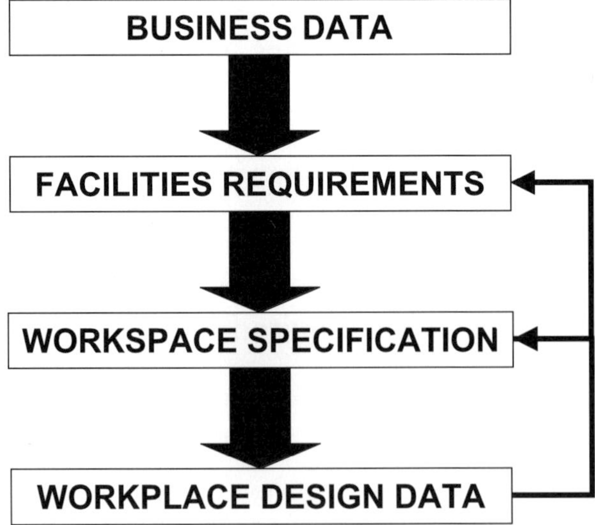

Fig. 4.1 The process of converting business data into workplace design data.

- The types of workplaces required, e.g. individual workplaces, collaborative, meetings, quiet, communicative, presentation, social, open and enclosed.
- The internal locations of workplaces, which will include the need to aggregate or disperse workplaces.
- The external locations of workplaces such as 'central' business locations, satellite premises, customers and suppliers premises, transit locations and the homes of people in the business.
- The duration for which the organization, and each of its constituent work groups, will require the workspace, noting any variations from one work group to another – again so as to be able to assess the impact one set of requirements may have upon another.
- Process operating environment and conditions – which may require segregation from, or co-location with, other work groups.

4.1.3 Services

- **IT infrastructure** – the infrastructure required to support the business, such as information technology, communications systems, LANs and WANs, telephone, fax, e-mail, document image processing and video conferencing.
- **Support services** – the support services for each work group and the business as a whole such as security, reprographics, catering and cleaning etc.
- **Operating hours** – the hours of operation of the business and its constituent parts – some parts of a business may have different work hours from others.
- **Equipment and machinery** – the equipment required for the work processes and its anticipated pattern of their use.
- **Information** – the types of information required to support business processes and the patterns of generation, storage, access, retrieval and sharing.
- **Storage** – the storage required for documents and other records, raw materials, finished goods and consumables in support of work processes and their anticipated pattern of use.

4.2 Synthesis of data

As can be seen from the above lists, almost all aspects should be assessed at work group level, and by the aggregation of the needs of each group, also at organizational level. To enable the facilities manager to develop a statement of the workspace requirements for the organization as a whole encompassing the above, he or she needs to gather information about the business requirements of each work group, the synthesis of which s/he can use to develop an accommodation plan. The data from which the facilities manager will derive the information above, will typically include the following.

4.2.1 Workspace

- **Location** – in relation to staff, labour pool and skills availability, customers, suppliers and competitors, transportation nodes and services, and other parts of the business.
- **Workplace settings** – to support different styles of working such as solitary, collaborative, interactive, small group, large group, quiet and communicative activities in support of each process, activity and task of every work group.
- **Internal environment** – the appropriate ambience of the workspace for particular operations and processes, its decoration and visual appearance and any specific maintenance or treatment it requires.

4.2.2 Organization

- **Numbers of people** – who will be required to support the operations of each work group and hence, by aggregation, the whole organization. This would include employees, staff from other sites of the business, agency and temporary staff, students, contractors, secondees, customers, suppliers, joint venture partners and others who will be working from the premises for some or all of the time the premises are available for use.
- **Organizational groupings** – the size and structure of the work groups and teams that comprise each business unit or work group, and how they relate one to another, including short-term and temporary groupings such as projects and task teams.
- **Hours of operation** – the working hours of all personnel and work groups (including fixed, variable, standard and 'seasonal' hours) and details of the services they may require through time.
- **Communication patterns** – the interactions that the people within each work group have with others (internal and external to the organization) and the extent to which (in terms of criticality, frequency and immediacy) these communications need to be face-to-face.
- **Visitors** – numbers, types and sizes of groups, timing and duration of stay, who they are visiting and what facilities they will require while on the premises.

4.2.3 Work processes

- **Types of work** – details of the processes, activities and tasks that will be carried out, and the types and numbers of people who will be involved in each.
- **Pattern of work** – the sequence and duration of the processes, activities and tasks, and how they relate to one another.
- **Service levels** – details of the levels of services required to support the processes, activities and tasks of each work group.
- **Dependencies** – the key factors that are critical to the success of the operations of each work group, and of the organization as a whole, without which they would either not function or be severely disrupted.

- **Inputs and outputs** – the key inputs that are critical to the success of the operations of each work group, as well as the organization as a whole, without which the organization or any part of it, would either not function or be severely disrupted – this would include both the tangible and the intangible, i.e. products and/or services and the criticality of the outputs for other work groups and the overall business.

4.2.4 Infrastructure

- **Resources** – the types of resources required to support the processes, activities and tasks of each work group – this would include information in electronic form, access to hardcopy records of different types, both 'live' and archived.
- **Tools** – the types of tools required to support the processes, activities and tasks of each work group in terms of plant, machinery, hand tools and equipment.
- **Utilities** – specific utility services required for each process in terms of electrical power, gas, water, drainage and special services such as piped chemicals.
- **Environmental services** – specific operating conditions for each process in terms of heating, cooling, air provision, quality and extraction, and humidity control.
- **Materials** – the types and quantities of materials (raw materials and consumables) required to support the activities of each work group and their means and frequency of delivery.

It is of vital importance in developing an accommodation plan that the organization takes due account of its needs for business continuity and disaster recovery. The loss of workspace or services as a result of a fire, flood or other major incident needs to be assessed and contingency plans developed, as referred to in Chapter 6. Therefore, while gathering data on the critical needs of the business and its dependencies, it is essential that the facilities manager also explores, individually with each work group, their needs for contingency workspace and services in the event of a major loss, which can then be assessed in terms of criticality for the business as a whole.

4.2.5 Common support facilities

In addition to specific workplace data, the facilities manager also needs to develop details of shared facilities that the organization requires to sustain its activities. Common support facilities are those that are provided to service the needs of all, or predominantly all, of the users of the building. Typically the space requirements of common support services could include the following.

- **Archives** – secure containment for centrally held records, sorting incoming and outgoing records, microfilming and document destruction.
- **Building services monitoring** – to enable the status and condition of primary environmental services to be monitored and controlled.
- **Catering** – beverages, vending, snacks, full meal services, entertaining and corporate hospitality services, which may include support accommodation of kitchen, wash-up areas, storage and related staff areas.

- **Conference** – meeting rooms of various types and sizes, equipped where required with presentation aids and possibly supported by ancillary accommodation such as projection rooms, furniture stores and cloakrooms (see Fig. 4.2).
- **Display facilities** – to publicize and promote the goods, services and people of the business, to staff and visitors to the building.
- **Emergency facilities** – disaster recovery facilities, emergency control room and related facilities, contingency and multi-purpose incident facilities.
- **Facility services** – cleaning equipment and consumable stores and storage for stationery, copy and printing paper and related materials, furniture and building products.
- **Goods entrance** – to handle bulk deliveries of consumables and raw materials, and also finished product dispatch.
- **Health and medical facilities** – health and safety briefing facilities, first aid facilities, rest rooms, examination rooms, doctors' consulting rooms, emergency treatment rooms.
- **Information technology equipment** – facilities for central and distributed processing equipment, maintenance and repair facilities, data media storage and related equipment storage.

Fig. 4.2 Presentation and audio visual aids are essential elements in collaborative working settings (Canon NEP, Sunderland).

- **Library** – storage of centrally held reference material in hard copy and in electronic forms, reading areas and facilities for the sorting of incoming and outgoing books.
- **Mail rooms** – sorting centres for incoming and outgoing mail, dispatch areas, scanning for suspect packages and their containment.
- **Materials stores** – to contain raw materials, consumables, part finished goods and finished goods ready for dispatch.
- **Media facilities** – press centres, television and radio broadcast support services.
- **Plantrooms** – for machinery and equipment to provide environmental services such as water storage, mechanical ventilation, air-conditioning, electrical switchgear (and possibly transformers), electricity generators, fire detection and extinguishing facilities.
- **Reception** – for greeting and recording visitors, for visitors waiting, cloakroom and toilet facilities, enquiries and for securely screening and holding unwanted visitors.
- **Recreational facilities** – sports (indoor and outdoor) and games facilities, changing and locker rooms, showers.
- **Reprographics** – local photocopying, bulk copying, printing, graphics production, art work and document publishing services.
- **Security** – control room providing facilities to monitor the condition of entrances and exits in the building, the monitoring of building plant, the status of any restricted accesses, fire and intruder alarm monitoring, issuing of identity passes, managing security alerts and incidents.
- **Social and welfare facilities** – function suite, meetings spaces, bar, stage, changing and locker rooms, and showers (see Fig. 4.3).
- **Telecommunications equipment** – frame room, central service equipment rooms, hubs, satellite dish and tracking equipment.
- **Training** – individual learning centres, small and large group facilities, computer training, telephone training, demonstrations and training in the use of special equipment.
- **Vehicle parking** – internal and external car parking for staff and visitors to the building, goods and other vehicle parking, loading and off-loading, which may include garaging, washing, service and repair facilities.
- **Waste disposal** – holding areas for waste until collected for removal from the premises, which could include recyclable and non-recyclable waste, secure and non-secure waste paper, kitchen waste, wet waste and drainage, scrap and hazardous waste materials.
- **Workshops** – repair and maintenance facilities and associated materials storage.

In addition to common support spaces, consideration should also be given to providing facilities that support the personal needs of staff rather than solely their work needs. As leading management expert Tom Peters says, during working hours businesses employ the whole person, not just a part of them (Peters and Waterman, 1992). While this may appear

(a)

(b)

Fig. 4.3 Social spaces are a vital component of the work environment. (a) British Airways, Newcastle, (b) British Telecom, Prospect West, Apsley.

to have the paternalistic overtones of a bygone Victorian era, evidence shows that where an organization subscribes to the principles expressed by Peters, then it is all the more important that the business addresses the needs of the *whole person* during this period. Some businesses – such as British Airways at their Waterside Headquarters at Heathrow,

IBM at Bedfont Lakes, both of which are in London, and SAS in Stockholm – have recognized this need and have provided a wide range of services such as shops, banks, and even dry cleaning and hairdressing facilities on site.

However, it is not only the requirement for support space that needs to be considered, the types of support service to be provided also need to be determined, as they will have an impact on the amount of workspace required and how it is best serviced (see Fig. 4.4), which will include consideration of issues such as the following.

- **Type of service to be provided** – general needs and special needs.
- **Hours of service** – standard and special.
- **Recharge arrangements** – to work groups and individuals, on what basis, i.e. actual use, per capita or in relation to the area of workspace occupied.
- **In-house or contract** – method of provision of facilities services.
- **Security** – standard and special needs, hours of cover and response times.

In gathering the data, whether at strategic or tactical levels, it is of vital importance to distinguish between the *wants* and the *needs* of the people in the business. Often interviewees are not aware that they are in fact expressing *wants* in response to questions about their work requirements, because they see this as an opportunity (and possibly the one and only opportunity they will ever have) to say their piece about what is required. The result can easily be a 'wish list', which is likely to be of little real value to the facilities manager in developing an understanding of the real needs of the business. As we have mentioned previously, and make no apologies for doing so again, the response from the facilities manager's customers can also be a reaction to the interviewees' lack of real understanding of the topic and of the possibilities that exist, which can easily lead to a re-statement of the current provision. Therefore, as discussed in Chapter 3, the use of appropriate data gathering tools can go a long way to assisting the facilities manager with the screening out of undesirable 'wish lists'. Add to this structured interviews using questionnaires that are constructed with relevant closed rather than open questions, and you stand a much improved prospect of capturing data of real value.

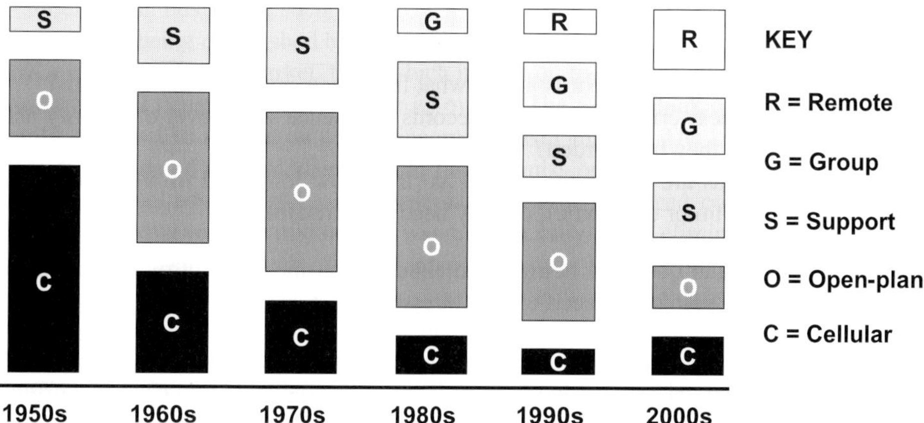

Fig. 4.4 The changing mix of workspace.

changes, and the degree to which such changes could impact upon the space required to accommodate hardcopy records.

4.4 Assessing the need for people

Information relating to the people in the business is of most significance in establishing the criteria for a business's workspace. This will include the number of people involved, the types of work they do and how they are organized. Of these, the most significant is the number of people who will be involved in the business and in what capacity.

Growing or contracting the business is not the same as growing or contracting the organization. In the comparatively recent past, analysis of most successful businesses would show a common thread, namely a direct, but often different, correlation between headcount and the financial success of the business. Be it car manufacture, grocery sales, banking transactions, or healthcare provision, the answer was always the same – successful businesses charted a rise in their headcount, while those whose business was in decline demonstrated corresponding reductions in headcount. This situation held good for many businesses until the late 1980s when Europe, North America and parts of the Far East experienced the worst economic recession they had seen for more than 50 years. Coincidental with businesses emergence from the recession, battle weary and scarred with its effects, was the explosive development in information technology and especially portable microelectronics – mobile telephones, computers and the Internet. These developments, together with the plummeting prices of computing power, both desktop and laptop, created an environment for dramatic change in the workplace. From the mid-1990s onwards, many very successful businesses were seen to be shedding significant numbers of employees. In other words, the correlation between business success and the numbers of people employed was broken in much the same way as it had been just over a hundred or so years previously, with the advent of what was to become known as the industrial revolution. Now organizations were reducing headcount not because they were in decline but while their business was growing (Fig. 4.5). In fact for some their very success was founded on their ability to change work processes to make them much less labour intensive.

Consequently, when gathering information on the future numbers of people the business may require, facilities managers need to explore the potential that exists for the introduction of new work processes, especially those that are likely to embrace emerging technologies, with a view to assessing their impact upon forecast personnel numbers. It is just such considerations that would feature in the scenario modelling referred to in the previous chapter.

Because facilities are the infrastructure that support the people in the business in achieving the goals of the organization, then clearly to ensure that adequate provision of facilities are made, the facilities manager needs to know the total number of people, or customers, whose needs he or she has to serve. First, and often foremost, this will be employees, but the facilities manager's customer base is far wider as, in most organizations, it will also embrace agency staff, temporary staff, contractors and sub-contractors, secondees and personnel from joint-venture and other business alliances. So the number of people the business requires to enable it to conduct its affairs is often substantially greater than at first it might

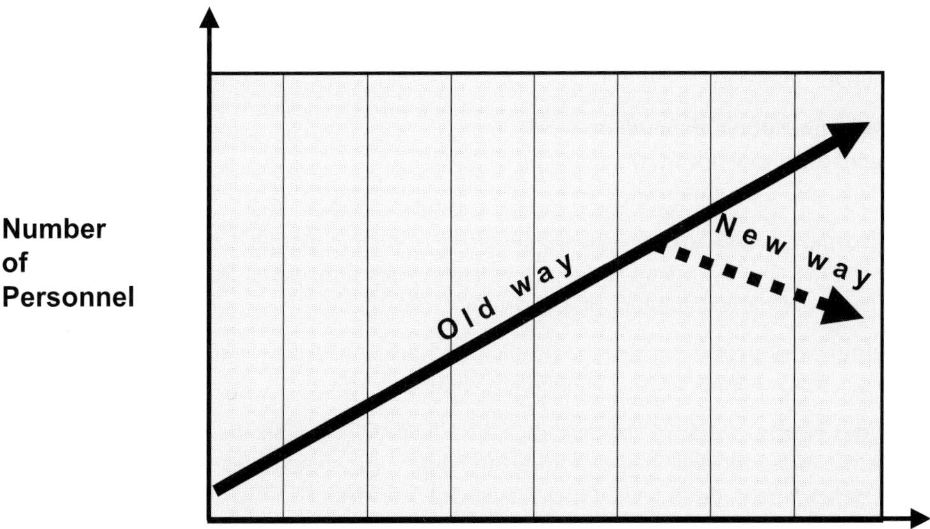

Fig. 4.5 The correlation between headcount and financial success is changing as businesses are able to do more work with fewer people.

appear. The facilities manager must therefore explore with business managers the current and future sources of labour that may be deployed in pursuit of their objectives.

Finally, when gathering information about the people in the business, the facilities manager must be respectful of the sensitivities that may arise. This is especially pertinent when the business is pursuing a course of organizational change which may affect the numbers of people to be retained in the business. Regrettably, there have been instances in some businesses where the announcement of redundancies, 'downsizing' programmes and mergers have been pre-empted by unfortunate 'leaks', which have caused much anxiety and distress. To avoid such occurrences, the facilities manager must ensure that all data gathered in connection with the future plans of the business are the subject of appropriate levels of confidentiality and security.

4.5 Assessing the shape of the organization

Following-on directly from the number and type of people required by the business, the facilities manager needs to know the way in which these people will be organized and grouped. Conventionally, most of the people in the organization would have been employees, and again conventionally they would have been grouped by function into departments. For some businesses this may continue to be the way they are organized, but such businesses will be fewer in number in the future. The reasons for the changes could be manifold, but are likely to fall into one or more of the following five key areas.

- **Process-based work group** – where personnel of different functional disciplines are grouped in relation to tasks and work activities on a long-term basis or permanently.

- **Project-based work group** – similar to process-based work groups but where the duration of the group is usually short-term or temporary due to the nature of the activities involved.
- **Geographically independent work group** – this involves people whose activities require them to work from a range of locations, some or all of which may be remote to the business's own premises.
- **Third-party work group** – where the group of people involved are not employees of the business but are contracted to the business via a third party, joint venture, strategic alliance, outsourced contract or sub-contract.
- **Virtual work group** – where the people in the group interact for all or most of the time via electronic means such as telephone, e-mail and video.

Charles Handy (1995, p. 70) describes the 'shamrock' organization, which he predicts will be the organizational shape of things to come – for some it has already arrived. The organization Handy envisages, which he named 'cloverleaf', is based upon a variety of ways of providing labour to the business, other than 'conventional' full-time employees. Such arrangements would include employees at the centre, with other forms of labour provided arranged around it like the leaves of a cloverleaf, hence its name. These 'others' would include part-time and piece-workers as well as self-employed and freelance as mentioned above.

Gathering information about the organization's shape and structure, the facilities manager needs to investigate with business managers the current organizational structure as well as the likelihood of issues arising, such as the following.

- **Different ways of engaging people** – methods other than via conventional employment can influence where people are located, i.e. within the business's core building or in remote locations such as contractor's premises or at home.
- **The structure of the organization** – a de-layered flat structure or one comprising self-managed teams can influence the size of workspace modules required.
- **User control** – with higher levels of empowerment in the organization, the facilities manager is likely to encounter higher levels of expectation from his customers who will demand greater control over their work environment.
- **Function or process** – a debate that has been going on for almost as long as space planning has been in existence, relates to whether people should be grouped in relation to their function within the business, i.e. finance, engineering, HR, or by the work process in which they participate – the issue is most commonly debated in organizations that practice matrix management.

Changes from functional groups to process oriented groups will pose challenges for the business, and for which facilities managers need to find workspace answers. Key drivers for the co-location of workplaces that are likely to be encountered are the following.

- The ability to form and reform work groups in response to the varying nature of activities, e.g. project teams.
- The ability to form and reform work groups in response to varying quantities of work, e.g. peaks and troughs of work flow.

- The ease with which the business is able to get the necessary participants together to resolve a problem, create a proposition or give directions when required.
- Co-location based upon the criticality of the need for face-to-face interaction, where criticality is defined as a function of frequency, immediacy and importance.

The innate capacity of the workplace to influence the behaviour of its users, can be used by the business to improve the morale and hence productivity of its workers (Fig. 4.6). Senior managers can use this attribute to great effect, when seeking to bring about major cultural change within the organization. Separating rank or status in the organization from the allocation of workspace; providing facilities and services based upon work needs and not upon position; and creating open rather than closeted workplaces, all of these can be used as instruments in the process of making cultural changes in the business. In considering the space needs of the organization, the facilities manager must be wholly conversant with the aims of senior management and their aspirations for how the workspace should reflect their desires for the cultural direction of the business.

4.5.1 Work and work processes

The primary purpose of many buildings is the provision of an environment that supports people and the work processes they carry out – hence the term *workspace*. It follows therefore that next to details about people, the most important data the facilities manager needs to gather relates to work and the processes it involves. But what is work? At a very

Fig. 4.6 The overall ambience of the workspace can contribute to workers' morale and hence their performance (WM Company, Edinburgh).

simplistic level work can be defined as the application of resources to the conversion of knowledge, skills and energy, into goods and services that are supplied to customers (Fig. 4.7). Many of these will be sold for profit, some will be provided free at the point of consumption.

In most work processes the workers will receive payment for their labours, in some they will give their time free, i.e. voluntarily. The common link that all work activities share is the need for a supportive environment, tools and services that are provided cost effectively. Therefore, to enable the facilities manager to provide the most appropriate environment commensurate with the business's objectives, he or she needs to have a solid understanding of the activities that will be carried out within the premises, i.e. the work processes, which will necessitate gathering data on the tasks people carry out, how they relate to one another, their dependencies and the tools, services and resources that are required.

Space standards can be developed from the space required to undertake satisfactorily particular job functions. The area required for each task or activity is built up from an analysis of the space required for its constituent parts, which are themselves derived from examination of the processes involved, research and experience. To be effective, space standards must be founded upon objective requirements and not upon impressions of needs and subjective desires – 'functional comfort, rather than square footage', as Jacqueline Vischer (1996) puts it. That means the facilities manager must understand the significance of the contributing factors that influence functional comfort, upon which the workplaces will be based. By close observation of existing working practices coupled with as wide a review as possible of different individuals undertaking the relevant tasks, the elemental

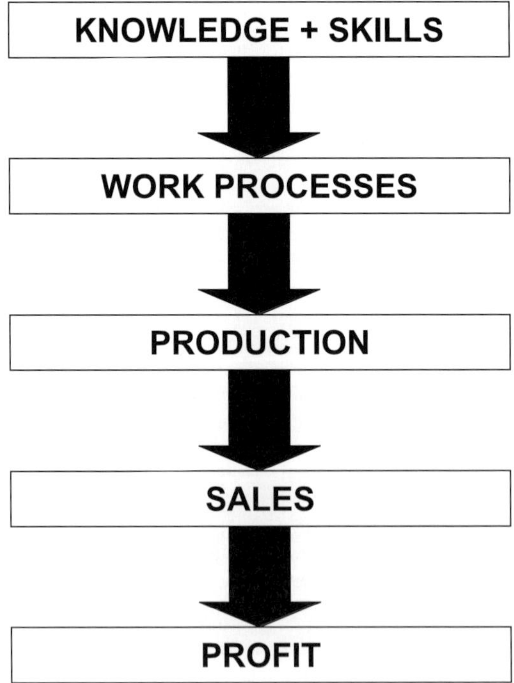

Fig. 4.7 The relationship between knowledge, skills and profit.

areas required can be determined, from which a given space standard can be developed. However, observation of existing work processes may not provide all of the information required for the future needs of the business. Work activities have changed in the past and will change in the future – even those that have changed only recently. Most, but not all, of the changes in work processes are a result of the introduction and development of technology. One only has to consider the impact that technology, especially the PC, has had on work processes in almost every place of work, within a very short span of time.

The key issues to consider are:

- **Changing patterns of work** – balancing the needs for solitary and collaborative activities;
- **Changing types of work** – where work was, for example, previously paper-based it may become increasingly electronic-based, or a combination of both;
- **Changing work processes** – where an increasing amount of technology is applied to previously manual and labour intensive activities, which are being replaced by automated or semi-automated processes.

When gathering information about the organization and its work processes, care needs to be taken when using job titles as some can be misleading and the same title can refer to quite different roles in different work groups. For example, until comparatively recently, a title such as 'Major Accounts Manager' would lead one to expect that it referred to someone who managed the relationship with large customers or accounts on behalf of a sales or manufacturing business, while others in the organization handled smaller customer accounts, which in turn suggests a form of working group. Today, some organizations, despite having larger volumes of business than in the past, have fewer customers. In the food manufacturing industry, there was a time when a large volume of business was synonymous with a correspondingly large number of customers. The customers could either be those who buy the products direct from the manufacturer through their own retail outlets, or from other retailers to whom the manufacturer wholesales their products. Today the geography of retailing has changed dramatically for many manufacturers whose principal, if not only, customers are the major food retailing chains and supermarkets, such as Tesco, Sainsburys and Asda. Therefore, the work patterns of a 'Major Accounts Manager' today are likely to be focused upon one customer with many outlets and hence the business of the manufacturer may be channelled through a very small number of customers. The work patterns of such 'Major Accounts Managers' are thus very different than they would have been in the comparatively recent past. Previously, the job would have entailed visiting many retail stores of several major customers. Therefore, although 'site visits' are still an integral part of their work activities, much of the business is conducted via telephone and electronic data interchange (EDI), where the role of the 'Major Accounts Manager' is primarily customer liaison and trouble-shooting rather than 'order taking' as previously would have been the case. Consequently, as this example highlights, it is essential when reviewing work processes that support for the business is founded upon a thorough understanding of what tasks and activities people actually carry out and not what is suggested or implied through their job titles.

Observation of people at work must, however, be conducted with care as for some it is synonymous with the time and motion studies that were much in evidence in the 1960s, when a man in an overall with a stopwatch and clipboard was, at that time, the cause of much industrial unrest.

Where the work processes are flexible in terms of timing, people and volumes, then it is likely that the work environment will also need to have some flexible capacity in terms of the numbers, workplaces and people it can accommodate, particularly in relation to flexible work settings.

4.5.2. Existing workspace

Reviewing the existing workspace of the business can be useful in helping to understand people's needs. Conducting the assessment with an open mind is as important as doing it with open eyes, to ensure what is obtained is of value, and not a distraction to the objectives in hand. Therefore, the types of questions that need to be asked as we view the workspace in-use would include the following.

- What do we see and what does it tell us about the workplaces and the users?
- Is what is being observed a typical pattern of use, if not why not?
- Is observation of the workplaces affecting its use, if so why? And is it unintentional or deliberate?
- To what extent is the existing workplace affecting the process of work, i.e. is the building constraining activities, flow patterns and the effectiveness of operations?

Further, by their actions in the way they use the workspace, what are the current users telling us about the following.

- How well the existing workspace and its infrastructure meet their needs?
- To what extent are workers making good use of the workplaces at their disposal?
- To what extent are workers using the workplaces in the manner that was expected and planned?
- And, if this is different to what was envisaged, why is it so?
- If some aspects of the work environment are being used more than others, why is this so?
- Do workers appear to be fully aware of all of the capabilities of their workplaces, if not would demonstration and training be beneficial?
- Are the workplaces being well looked after and maintained, and what does this tell us about the level of servicing being provided, i.e. is it adequate, over or under the required level?
- What are the occupancy durations and patterns of use for each type of workplace provided, and what do these tell us about the levels of provision and the capabilities of each workplace type?
- What aspects of the workplaces are being well used and what are not being well used, and why?

Seeking information about the business's workspace requirements, particularly by viewing existing workspace, can lead to presupposing that the business needs, and will continue to need, workspace. As previously mentioned, the facilities manager must always be alert to the fact that an answer will always be influenced and 'shaped' by the question and the way it is asked. Hence, when seeking to determine the criteria that will effect the amount of space required by the business, the facilities manager must ensure that the first question asked is – does the organization need workspace? Clearly, if the question was posed in this direct fashion then most business managers would not only respond in the affirmative, but would also then question the state of mind of the facilities manager. As previously stated, in the absence of any guidance about other options that may be available, the most likely response will be a re-statement of the status quo, i.e. we have a building, therefore we must need a building in the future. But is this true? And will it always be so? Clearly there are businesses that operate successfully without business premises, where 'employees' work from temporary locations such as customers' premises or transit locations such as hotels, or from their homes. Therefore, even if a work group or the whole business has been used to working from dedicated premises, there may be more effective, efficient and appropriate ways to work in the future which rely less on dedicated premises or which do not require any at all (Fig. 4.8).

In assessing the requirements for workspace, the facilities manager should therefore be asking questions such as the following.

- Is the business's need for workspace constant or variable?
- Upon what is the amount of workspace dependent?
- Does all of the workspace need to be aggregated together in one location?
- If it does, will it always need to be aggregated together?
- If not, what are the drivers and their timing that could influence the aggregation or the dispersal of workspace?

Fig. 4.8 Technology now makes it possible to work almost anywhere (source: BT Corporate Picture Library; a BT photograph).

4.6 Workspace as a tool

The Oxford English Dictionary defines a tool as – 'a means of affecting something'. Surely this definition applies to the place of work. Work buildings are places to accommodate people, processes and tools which, when they are brought together in an appropriate way, enable organizations to operate businesses. However, in addition to a workspace being the 'container' of the people, processes and tools of a business, it should also be considered as a tool in its own right.

For some businesses the identification of their premises as a specific tool or series of tools is more evident than others. A supermarket, cinema, an ice rink and a bank branch are all evidently tools in that they *shape* the product or service that is provided and, in fact, some – such as a cinema or sports arena – are obviously integral parts of the 'product' that the business is selling, i.e. the *experience* of viewing films or watching sport. Similarly with factories, hotels and hospitals that supply the means, i.e. the environment and its infrastructure, by which the product or service is created or provided – in other words a *tool* by which its users create their product. The office is no different, or at least it should not be considered any differently. It too helps its users to shape the products or services they are in business to provide. It is well known that a poor work environment can have a detrimental effect upon the productivity of the people that work in it. How much better then to provide an environment that supports and encourages the people in the business in their endeavours, rather than impeding them. Hence, the facilities manager determining the workspace needs of the business should be assessing how the premises that are provided, or could be provided, could help the business make its product more effectively. In exploring this aspect, consideration needs to be given to the following.

- Where is work to be carried out?
- Possibly a more relevant question would be, where *should* work be carried out?
- Does the organization need a specific building, or building type, to conduct its business, or could the work be carried out in a range of building types?
- To what extent have conventions dictated the selection of business premises in the past, and to what extent should they influence selection in the future?
- Would different premises enable the business to produce its products or services more effectively and efficiently?
- How easy will it be for the business to dispose of the building, if or when it is no longer required?
- Is the building too tightly tailored to the old or former needs of the business, making it unlikely that it will suit the future needs of the business or other possible occupiers?
- Can the building be sub-divided to enable discrete parts to be released for use by other parties?
- As occupation levels in the building increase, then it is reasonable to expect that there could be more cars at the building than previously – can the car park provision cope with the increase in demand and the segregation of the vehicles of different occupiers?

4.7 Developments in technology

The single most significant contribution to changes in work, work processes and consequently the workplace throughout the 1980s and 1990s, if not in the entire history of work environments, has been as a result of developments in technology in general, and information technology and PCs in particular. Many businesses did not appreciate, and hence anticipate, the scale and timing of such developments let alone the impact they would have upon their people and their workspace – and this despite the efforts of Frank Duffy and others to forecast the impact and guide possible solutions (Duffy, 1983).

In gathering information about the future direction of their business, facilities managers therefore need to investigate planned as well as possible developments in technology and the likely demands they will place on the work environment and the implications for the amount, type and location of the premises required. The development of technologies such as the fax and the mobile telephone, very rapidly became absorbed into the workings of business and, in fact, have become an integral part of our daily lives. Even apparently very sophisticated technologies such as the Internet and video conferencing are rapidly changing the way business is conducted. Therefore, the facilities manager must, in conjunction with his IT colleagues, assess the likely demand in the business for new technologies and their impact upon work processes. The principal technology developments that are most likely to impact upon work processes over the first 5 to 10 years of the 21st century, with consequential effects for the amount and type of workspace a business will require, will emanate from the following.

- **Mobile phones** – the global population of mobile phones reached almost 150 million in 1997 which represented a tenfold increase over the preceding six year period and is expected to double by 2000. This will continue the expansion of mobility of working practices.

- **Flat screens** – the development of flat visual display screens as a component of laptop computers, such as LCD and plasma panels, will progressively make its way to the desktop environment with great impact, reducing the space required by a CRT monitor by 60% as well as generating much less heat output, significantly reducing the prospect of stray reflections and glare on the screens while improving the quality of the projected image.

- **Wireless LANs** – although at present performing at around 20% of equivalent wired systems, as the performance gap narrows and set-up costs reduce, the ability to create networks without the use of cabling may lead eventually to the demise of the almost ubiquitous raised access floor, with resultant reductions in building costs.

- **Video conferencing** – the much heralded expansion of video conferencing has been slower to materialize than many techno-pundits predicted in the 1980s. However, there is evidence of a growing pace of acceptance with the advent of low cost videophones and desktop computer video systems. Similar to other innovations, as the quality of the technology (video image, audio synchronization and transmission speeds) improves to a standard equivalent to television, then the market penetration of these products (for both business and personal use) will rapidly increase. Most products today include not only face-to-face audio and video, but also 'data-conferencing', which allows two parties simultaneously to share the same application and to work together interactively in real time, despite being located geographically distant.

Add to these innovations in technology the unrelenting expansion of software applications enabling more activities to be conducted by fewer people in a faster timescale, then it is apparent that work, work processes and the work environment are bound for a turbulent and exciting future. Specific future developments will of course be difficult if not impossible to predict, but the facilities manager should always be conscious of the need to prepare for developments that are likely to impact upon the business and its requirement for, and use of, workspace. Through consultation with appropriate managers in the business and experts outside the business the facilities manager should strive to keep him or herself appraised of trends in technology development. The facilities manager should be particularly alert to the implications for factors such as the following.

- **De-centralization of workspace** – changes in the numbers of people that need to be aggregated together, as technology will enable them to carry out business operations at dispersed locations.
- **Location of workspace** – the use of wide area networks (WANs), the Internet and video conferencing technologies makes an expansive geography available, permitting people to work from a variety of locations.
- **Amount of workspace** – in addition to the follow-on from locational factors which could change the amount of workspace needed in particular locations, the overall amount of workspace required could change because fewer people are required to carry out the same volume of business.
- **Workplace size and layout** – developments in technology have had, and almost certainly will continue to have, an impact upon the size, shape and configuration of workplaces, e.g. the introduction of flat screen monitors in place of CRT monitors will reduce the amount of worksurface required and change the shape of the workplace and the way it is clustered.
- **Workplace environment** – covering issues such as heat generation from increased densities of equipment, such as in call centres, and hence placing greater demands on HVAC systems; quality, quantity and control of lighting to address the needs of technologies such as video conferencing systems; control of ambient noise levels resulting from increased use of communication systems.
- **The allocation of workplaces** – as technology enables more tasks to be carried out by fewer people, then the workplace settings that are at the disposal of each individual will need to change so that people can make use of a range of settings that are appropriate to their tasks, rather than being compromised in one workplace.
- **Amount of ancillary workspace** – a change in the amount of ancillary accommodation that the business requires resulting from the adoption of technology, e.g. a reduction in hardcopy document storage space required is one benefit derived from the use of document image processing (also known as work flow management technology) and the use of EDI.

Case Note 4a

Insurance company, Southern Holland. At its headquarters, where flexible styles of working are practised, this Dutch insurance company has equipped the landscaped grounds around their office buildings with data points, which enable staff to work outside by plugging their laptop computers into the local area network, giving them the ability to work out in the open, weather permitting!

Case Note 4b

Computer software company, Southern England. This company provides a virtual shopping link with a nearby supermarket, where staff can order their groceries via the Internet, which are then boxed and delivered to a pick-up point for the staff to collect on their way home.

Case Note 4c

Science Agency, Southern England. Here science staff consider the direction of, and funding for, fundamental research. The people are based in their own individual and very 'monastic' cellular workplaces; this reinforces the concept of isolationism and hence misses an objective of the Agency, which is to encourage in the businesses in whom they are investing, the demonstrable merits of collaboration and cross fertilization of ideas. Clearly they were not fostering this in their own style of working. Those involved were themselves very senior and eminent scientists who had come from industry and had been used to such cloistered environments, but in their current organization the setting should have been more supportive of interaction and sharing, than of solitude and the sole possession of information.

Case Note 4d

BP Exploration, Aberdeen, Scotland. The term 'outsourcing' is one which will be well known if not well experienced by most organizations. In 1992, BP Exploration, based in Aberdeen, helped redefine what most people considered as being appropriate non-core activities to outsource, when they contracted-out their finance function, which at that time employed around 250 people, to accountants Arthur Andersen. Not only did this have profound effects on the day-to-day operations of BP Exploration, and of course of the people concerned, but it had major implications for the BP facilities managers who almost overnight, had a void in their workspace where previously 250 staff had worked. For many organizations this would have created many difficulties, even if the 'extra' space would have helped ease congested workplace layouts – too much space, consuming resources and incurring costs. Fortunately for BP Exploration, they were simultaneously attempting to rationalize their accommodation following their earlier acquisition of Britoil, who had been occupying some 20,000 sqm of accommodation elsewhere in Aberdeen. Therefore, by re-organizing their departments' locations, they were able to provide Arthur Andersen with a ready-made office in part of the former Britoil premises, which BP were eventually able to vacate.

4.8 The geography of the workplace

As we have shown, work has changed and is continuing to change, through the use of new tools, increased skill levels, and the development of new processes. As work changes and individuals take on a wider spread of activities and a wider span of responsibilities then it follows that their requirements for a workplace also change. If, in the past, it has been accepted that a 'team leader' needs a workplace to meet his or her needs, which has different attributes from that of an administrator, or a VDU operator, then when the roles of people change to embrace some or all of these tasks, is it not sensible to expect that the individual will require different workplace settings to address the needs of those activities that were previously carried out by several people? The demise of the compromised conventional workplace has, as Veldon and Piepers (1995) explain, through its failure to reconcile the conflicting demands of many of these tasks, led to the need for an approach that enables individuals to have a range of environments from which to choose, in order to meet their specific task needs at any point in time. In short, if the correct tools are required by someone who is doing one job or task continuously, then surely it follows that they are no less necessary for someone who carries out these tasks as *part* of their work. To reject this proposition is to suggest that because the task is only performed on a Thursday or only between the hours of 10.00 am and 12.00 pm each day, providing the correct 'tools' for the job cannot be justified. This cannot be correct, as we all need the appropriate tools for the job. This is a subject to which we will return in Chapter 6.

The role of the workplace is to support work not to constrain it, hence it must be capable of adaptation through time in response to the 'flexing' of the business and its requirements. Therefore, in gathering information about the internal geography of business premises, the facilities manager needs to assess the following.

- Increasingly, the need for co-location of people will be based upon their need for face-to-face interaction rather than for the sharing of data, which previously could only be achieved through physical transfer of data, but today can be achieved by electronic systems deployed from remote locations;
- In establishing the levels of interaction, and hence the need for co-location of working groups, it is important to recognize that in all but a few rare instances, it is most likely that the relationship will not involve all of a group or department with all of another group or department, but rather the interaction will occur between some of the personnel in each group. Further, the numbers of people involved in the interaction may change through time;
- Sharing department or working group resources will lead progressively to a 'fuzzing' of the boundaries between work groups. In time this fuzzing may give way to an overlap of territory, as indicated in Fig. 4.9, which will create the potential for significant savings in workspace and other space related resources. When considered for the business as a whole, and as promulgated by the likes of Charles Handy, this is likely to result in the demise of functional silos which have for so long been a characteristic of most organizations.

What is clearly not going to continue to be acceptable, if it ever was, is the practice of providing individual staff who, at various times, work from different company locations,

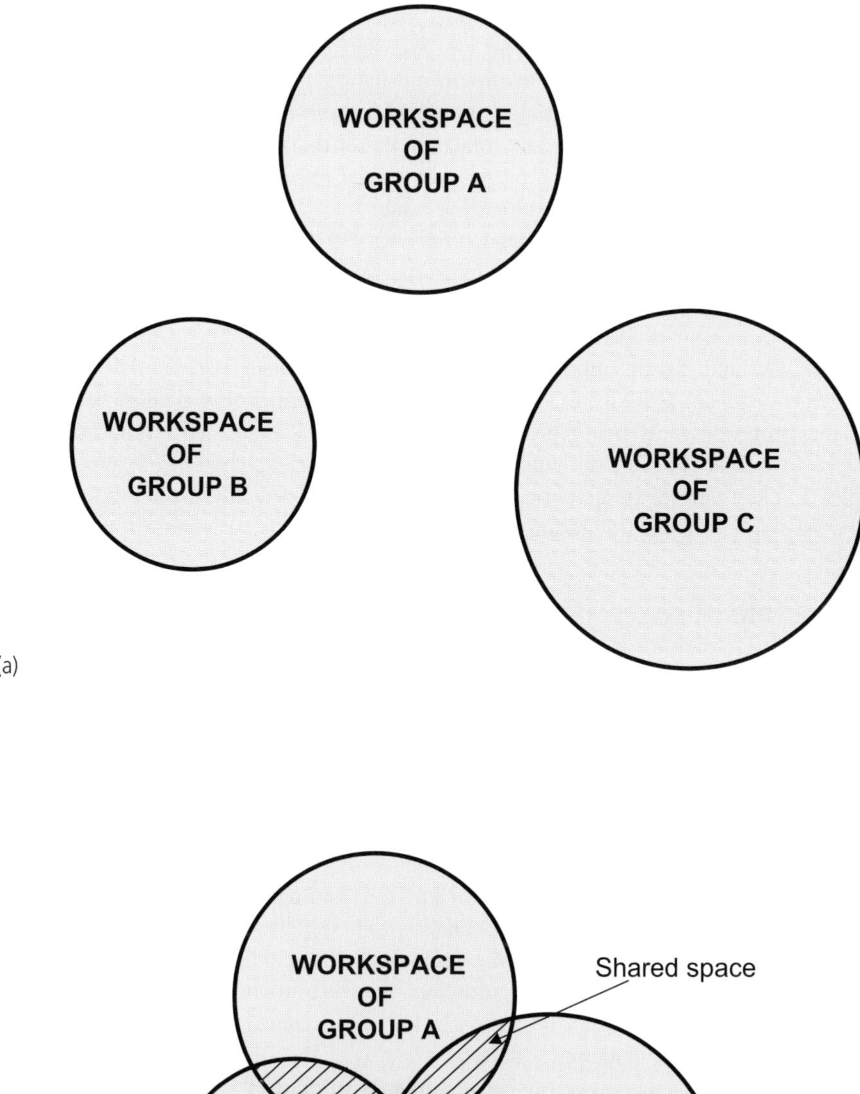

Fig. 4.9 (a) and (b) The sharing of workplaces and support space between several groups can have significant benefit in terms of reducing the overall quantity of workspace required by the business – facilities shared by all work groups could be provided by means of an 'internal business centre'.

with workplaces in each location. For example, staff of an insurance company, who are based in their London head office, and who are required for possibly 2 to 5 days per month (and sometimes less) to work from the company's principal processing centre located over 60 miles away, are used to having 'their own desk' in both premises. This is a waste of corporate resources, founded upon some mistaken belief that the provision of workspace is somehow commensurate with rank or an individual's importance in the enterprise, or simply that there is no other way to address the issue.

More prevalent in these times is a visitor making use of someone else's temporarily vacated desk when visiting the premises. While superficially this may seem to be a prudent use of facilities, it is only so if the environment and, particularly, the workplace meet the needs of the temporary occupant, *and* the combined occupation of the workplace, i.e. by predominant user and visitor, optimizes the use of the workplace. If it is otherwise, then the temporary occupant is likely to have his or her work needs compromised because of poor or inappropriate tools being provided, as referred to above. This type of use of workplaces can mask the poor utilization of business resources. Therefore, as mentioned in Chapter 3, a key part of the gathering of data about the organization's needs should be a detailed assessment of the utilization of workplaces.

4.8.1 The role of space standards

As will be discussed in Chapter 6, the development and use of workplace space standards can have many benefits. However, in assessing the business demand for workspace, the facilities manager should be mindful to question the continued relevance and application of existing space standards. That is not to say that space standards *per se* are not relevant, but rather that the facilities manager should question whether the existing standards are still relevant in the light of current rather than previous working practices. The trap that awaits the unsuspecting facilities manager and space planner, is the blind application of standards because 'that is what is laid down in the planning manual'. Space standards cannot be finite or definitive as they are dependent upon the hardware included within the footprint, i.e. the furniture products and equipment, which will change through time. Therefore, standards should be used as guidelines for the amount of workspace required for particular tasks and should not be slavishly followed where work processes or work tools render them obsolete or inappropriate.

To assess the ongoing relevance of space standards so that only appropriate ones may be relied upon in developing the workspace for the business, the facilities manager should give consideration to the following (see Fig. 4.10).

- **Range of workplaces** – the recognition that most people are not carrying out single, or mono-tasks, but are many or multitasking and hence need a range of workplaces to be at their disposal, e.g. meetings spaces, service spaces, spaces to support collaborative and quiet working.
- **Taskplace settings** – identification of the critical components required to support specific types of work activity.
- **Combined workplaces** – the degree to which the workplace needs of individuals and teams is compromised by attempts at combining the requirements of more than one task within a single workplace.

(a)

(b)

Fig. 4.10 (a) and (b) Interactive and collaborative styles of working require responsive and supportive work settings.

- **Benchmark space provision and use** – through reliable comparison with other businesses, assess the organization's current and proposed workplace allocation, provision and utilization with a view to adopting best practice.
- **Flexibility** – review space standards in relation to their ability to address the changing needs of the business in terms of technology, buildings and workplace products and hence the business's vulnerability to the impact of such changes.

The allocation of workspace has often been made by job grade, job function or a combination of both. The critical basis for allocating workspace must be by job function, i.e. the area provided should be sufficient to permit the tasks to be carried out and hence be capable of supporting the required workplace hardware. But just as important as the amount of space is the *quality* of space provided. In fact, we believe that if the quality of workspace is right then the quantity, although still important, becomes less critical for effective working.

Inevitably, in all but the most democratic of organizations, an element of status features in the allocation of space. In their study of UK workplace densities, Gerald Eve (1997, p. 9) identified that 44% of businesses have no formal policy for allocating space, but do so by staff grade and status. However, we are pleased to note that our experience indicates that this practice is on the decline as businesses recognize that there are other more effective and potentially less costly ways of recognizing status, where its provision is an inescapable trapping of management.

In assessing the amount of workspace that is required by the business, the facilities manager needs to be very wary of using 'rule of thumb' allocations, such as 10 sqm per person, where the area has been determined by some rough and ready method, most probably an overall average for the business. Such methods should be avoided for all but the broadest of planning purposes, since their application is dependent upon understanding the underlying assumptions upon which the 'rule of thumb' figures were determined, which frequently are not known by those whose use the 'data' – what we call *blind benchmarking*.

In addition to legislative requirements for fire escape etc, the need for circulation space has to be considered in relation to the number of people it serves, including adequate and appropriate access for disabled persons, as well as the widths needed for different types of access, such as trolleys for deliveries, servicing and maintenance.

As has already been mentioned, consideration also needs to be given to amenity space. However, in addition to the provision of general amenity space, specific attention needs to be given to areas where the business is likely to have a high density of workplaces, such as call centres. In such situations the amenity space can be seen as a trade-off by way of compensation for those who work at call centre activities and/or participate in workplace sharing. The absence of 'break-out' spaces has been seen to be a contributing factor in poorly designed call centres where attrition rates of over 30% per annum are typical. Such dramatic effects upon the business are felt in many ways – cost and productivity being the most obvious. An improved ambience can be achieved by enhancing the secondary circulation space to reduce the overall density of the workspace, as well as by the provision of break-out and social spaces. Consequently, the facilities manager needs to make provision for such allowances, in addition to the requirements for 'traditional' common support spaces, when assessing workplace requirements.

However, the best researched and most effective space standards are of little value if the space is empty or under-used. Therefore, the development of taskplace guidelines should always be set in the context of appropriate workspace allocation leading to effective utilization, i.e. workplaces that are developed taking account of the consumption of workspaces and their use through time, a subject we consider in Chapter 6.

Case Note 4e

Financial institution, Southern England. As an integral part of the development of an accommodation strategy for its head office, a review of the current use of its premises highlighted that, in a building that accommodates approximately 1200 workplaces, routinely 200 were unoccupied at any one time, and on occasions this rose to over 350 workplaces. The annual price of this under-utilization of workspace was assessed in terms of premises running costs as being in excess of £2 million.

Case Note 4f

Insurance company, London. A similar review of this business's accommodation, supporting 1200 people, identified that meetings spaces, both enclosed and open, have typical utilization levels of 25% or, expressed differently, the meetings spaces are on average unused for 75% of the time. Yet frequently people in the business were unable to find spaces to hold meetings. Investigation identified the principal reasons for this situation as the following.

- **Meetings spaces in cellular offices** – almost all enclosed workplaces incorporated a meeting table, and these tables, when not being used by the occupant of the office, were not used by others due to the potential for distraction for both the occupant and those using the meeting table.
- **Forward planning of meetings** – to avoid having difficulties in obtaining a meeting room at short notice many staff would book a room up to a year in advance and to 'block book' rooms in anticipation of, but without the certainty of, needing the room, which frequently resulted in rooms being booked that were not subsequently required.
- **Control of meetings spaces** – many meeting rooms were under the direct control of specific work groups with no central administration, and as a result there was no method whereby the room's availability could be 'broadcast' to all potential users, and hence the availability was unknown, leading to the belief by some that there was a shortage of places to hold meetings.
- **Other uses** – some meeting tables were identified that were almost permanently covered in papers, indicating that their principal use was not for meetings but was as layout or reference space. In some instances the papers did not appear to be disturbed for prolonged periods, suggesting that the tables were in effect acting as storage spaces.

4.9 Forecasting future needs

The road to hell is paved with good intentions, or so it is said. This could also apply to well intentioned facilities managers attempting to forecast the future workspace needs of their businesses. The reality is that the facilities manager is no clairvoyant and there are no crystal balls to help him or her. In this context an advertising campaign springs to mind in which a British university college, seeking to solicit new students, extolled 'we cannot predict the future, but we can prepare for it'. This too should be the battle cry of facilities managers as they seek to assess the future workspace needs of their businesses.

Much the easiest of data to collect, although at times it may not appear to be so, is that which relates to the current needs of the organization. Usually the current needs are either being effectively met by the prevailing provision or the shortcomings are well known to all! However, when seeking data about future needs the collection process is much more problematic. The first trap that awaits the unsuspecting is that frequently what is advised by interviewees as being the future need is a re-statement of the current provision. Usually this is for one of the following four reasons.

1. First let us eliminate the obvious – it may be that what is stated is exactly what is required, that is to say the current provision is assessed as being adequate for future needs, in which case all is well. However, there are possible reasons why this may not be the case.
2. As mentioned in Chapter 3, workspace and facilities are second nature to the facilities manager and his or her team, but to most business managers they are unfamiliar territory with an unfamiliar language. Consequently, they may not be able to articulate their needs and so the second reason for requesting the status quo is that the business manager may be unaware of other available options and hence restates the current provision in the absence of knowledge of any other proposition.
3. The third reason is distraction and disinterest. For most business managers the process of providing information about the needs of their work group many months and sometimes years ahead is an apparent irrelevance because their focus is on achieving this month's, quarter's or year's targets. If these targets are not achieved there will be no need in the years ahead or none that will concern the manager! So being asked, as some would see it, 'a bunch of dumb fool questions' about future numbers of people, desks, offices, storage etc, is a distraction to the manager from meeting his or her needs for the here and now.
4. This approach can be further compounded by business managers' belief that the facilities manager does not need to know this information now as 'we will deal with it when the need arises'. This is often said along with 'we have always managed that way in the past, so why do we need to change now?'. Hence giving rise to reason 4; namely, the mistaken belief that because the facilities manager and his team have in the past been satisfactorily able to meet the workspace needs of the group with information received at the last minute, there is therefore no need or benefit to plan ahead.

Whichever reason is applicable, what can be quite disconcerting for the facilities manager is that the business manger or customer representative will make their statement of requirements sound so authoritative. Therefore, for each of the above reasons, the facilities manager may have difficulty in realizing that the information he or she is being given is not as reliable as it may sound, or as one would wish.

A key component of forecasting future needs, is the necessity to establish not just that there is a requirement but over what timescales it is needed, and hence the facilities manager needs to address the following questions when assessing all future needs.

- How precise is the need expressed?
- How reliable have previous forecasts been?
- Is the need permanent or temporary?
- If temporary, when does the need arise and for what durations is it required?
- Is the need progressive, i.e. increasing through time and if so at what rate, or is it all required at a set point in time?

These questions apply to all aspects of forecasting businesses' work needs, but they are especially relevant to changes in headcount. For example, if a business unit forecasts that their personnel numbers will increase from the current 50 to 100 by the end of next year, does this mean that in late December next year an additional 50 personnel will be recruited, or that the personnel numbers will increase steadily from now until the end of next year, by which time the total will be 100? Or some other pattern? Therefore, when assessing future requirements, one should always be mindful to consider:

- What is it that people need?
- Why do they need it?
- When do they need it? and
- For how long do they need it?

The facilities manager should model workspace scenarios to assess the influences and impacts that changes to forecasts will have and then keep them under review. It should also be borne in mind that when assessing the future needs of the business the facilities manager should not assume that business managers are the founts of all knowledge. They too may not be aware of potential change, or be unwilling to accept it, in which case they are a poor source of data. The facilities manager must therefore also be alert to other possible sources of information, such as that which can be gleaned from others in the marketplace, about work process changes, new and different channels to market, technology development and life styles. All of these and more can be drivers for change in future workspace requirements. The speed at which the business will recognize the potential of such changes and accept them is what the facilities manager must focus his or her forecasting upon.

Having discussed how the demand for business accommodation is determined, in Chapter 5 we examine the other side of the workspace equation – the supply (buildings and their infrastructure) – before moving on, in the subsequent chapter, to consider how the demand can be reconciled with supply.

Case Note 4g

Financial services company, Southern England. The company technology development personnel, some 300 in number, whose task it is to develop the IT systems that enable the business to service its customers, and hence provide the tools used by most of the other staff in the business, were the people who had the least technologically appropriate workplaces in terms of size, layout and servicing. This work group were usually the last to have their workplace layout needs addressed and were generally driven to taking the 'law into their own hands' by making changes to workplace configurations that were not sanctioned by the facilities manager. The situation became so bad that progressively IT staff left the business in despair at what they saw as the inadequate support they were being given by the business. Only then, i.e. when many had departed, did the business take notice and appreciate that the shortcomings of the workplaces had to be addressed.

Case Note 4h

International chemical company, Northern England. As this business changed its focus, which resulted in the sale of many of its constituent parts, the company gradually appreciated that the specialist buildings and highly serviced environments that had previously been shared by all business units would potentially no longer be viable, which in turn would have disastrous consequences for the value of the assets it owned. The continued existence of the technical centre, and hence the asset value, was therefore dependent upon retention of the new businesses as occupiers. However, a building that had been built as a centre of excellence for a single occupying business had to address wholly new requirements as a multi-tenanted building. Most significant were the subdivision of services and the segregation of workspace. However, enhanced security provisions also had to be made, the need arising as a result of competing businesses occupying the same building campus. In other words, realizing the long-term benefit of configuring buildings flexibly for more than one occupier, requires consideration of more than just the physical segregation of workspace and services.

References

Duffy, F. (1983) *ORBIT Study* (London: DEGW).

Gerald Eve (1997) *Overcrowded, Under-utilised or Just Right* (London: Gerald Eve) p. 9.

Handy, C. (1995) *The Age of Unreason* (London: Arrow Business Books) p. 70.

Peters, T. J. and Waterman, R. H. (1982) *In Search of Excellence* (New York: Harper and Row).

Veldon, E. and Piepers, B. (1995) *The Demise of the Office* (Rotterdam: Uitgeverij).

Vischer, J. C. (1996) *Workspace Strategies: Environment as a Tool for Work* (New York: Chapman and Hall).

5

Assessing supply – the premises audit

In the previous chapter we considered the demand side of the space planning equation. We reviewed the wide-ranging demand variables that influence an organization's need for workspace. In this chapter, we will consider the supply side of the space planning equation. The supply variables are concerned with factors that influence decisions about the provision of the real estate resource and the servicing of workspace in support of business processes. The supply variables that are embraced within the premises audit, cover issues relating to attributes of the location of the building, its condition, its effective utilization and its facilities support infrastructure.

5.1 The premises audit

The objective of a premises audit is to assess how well a building is currently meeting the needs of the occupying business, and the degree to which it could satisfy future needs (see Fig. 5.1). Many of the issues are similar to those that would need to be undertaken as part of the process of evaluating new premises. The key issues of a premises audit will encompass:

- Location;
- Condition;
- Utilization; and
- Value for money.

Effective utilization of the premises is determined by the suitability of the environment for the work tasks to be performed. Therefore, in constructing procedures for the conduct of premises audits, it is essential that the needs of the business are known as they will act as the backdrop against which the evaluation process will be conducted. Consequently, the audit will – in addition to assessing the fabric and services of the building – also take account of how the premises are used. Analysis of the audit data will enable the facilities manager to identify aspects requiring attention, such as the maintenance of the building's fabric, improvement of services and the optimization of workspace layouts.

Clearly the single most important aspect of the premises audit is to identify what requires to be addressed to ensure that the occupier is receiving the best value for money that they can from the premises. This will include assessments of energy efficiency,

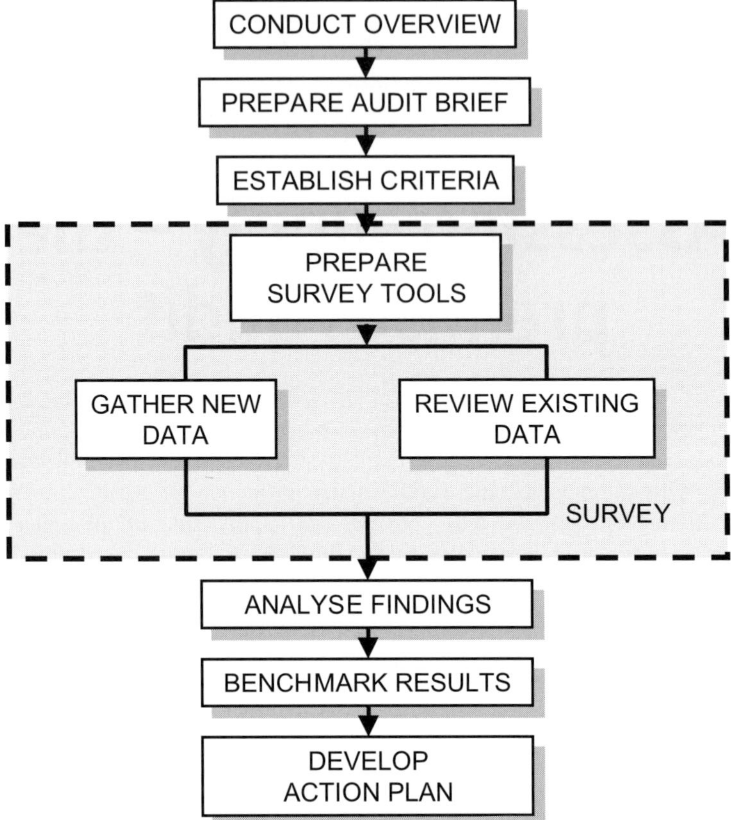

Fig. 5.1 The stages of the premises audit process.

adaptability and cost in use, bringing together all of the strands of the audit, i.e. location, condition and utilization. Some cost-in-use studies may be more easily carried out than others. For example, an assessment of the premises energy running costs would not be too difficult to undertake, drawing the information from fuel bills and occupancy periods. Assessing value for money in adaptation and internal changes, may be more difficult to achieve. However, in organizations with high levels of churn, the costs associated with premises adaptation may rival some fuel bills, and therefore have a significant impact upon facilities expenditure. The audit should therefore include an assessment of the costs of implementing workplace changes, be they partition moves, data/communications cabling re-configurations or others, and assess them against the provisions made for flexibility, i.e. the provision of, and investment in, relocatable partitions, raised floors and the like.

As with all audits, the premises audit only has a value to the organization if the investigations are applied as learning and feedback for the benefit of future premises operations. Typically we would expect a full audit, such as that described here, to be undertaken every five years. However, it is likely that some aspects, such as a review of the utilization of the premises, may be undertaken more frequently, possibly on an annual basis where an ongoing programme of churn and subsequent post-occupancy evaluation can be put in place.

In the main, facilities managers charged with the role of facilities provision in response to the organization's operational demands, will need information to address the following questions:

- **Where** – the siting of facilities relative to the location of labour, customers and suppliers.
- **When** – the timing of the need in relation to strategic business plans.
- **What** – the appropriate type of facilities, their attributes, range and profile.
- **How** – the optimization of the best fit between supply and demand over time.

In the past, the task of facilities provision was almost an afterthought of the business planning process. A study conducted in the UK that surveyed successful private sector organizations identified that this practice is not an uncommon occurrence (Then, 1996). One financial services company acquired a city centre office block and spent several million pounds refurbishing it, only to discard it as surplus and inappropriate to its requirements a year later. Another placed an unrealistic target on its facilities manager to deliver, within six months, a fully-serviced call centre to accommodate 200 new employees, in a city that was facing a shortage of office accommodation. In this particular case, a warehouse was converted in time to meet the corporate demand, but at a higher, premium cost than would have been incurred had a reasonable timescale been permitted.

These examples are typical of situations in which there is no informed dialogue between strategic corporate planners and business units managers on the one hand – (the source of demand), and the parties charged with the procurement and delivery of the workspace (the facilities team).

Figure 5.2 illustrates the communication gap in many organizations that is typified by the symptoms of under-management of the corporate real estate resource. The combination of a lack of clear strategic guidelines on real estate and facilities services issues, and

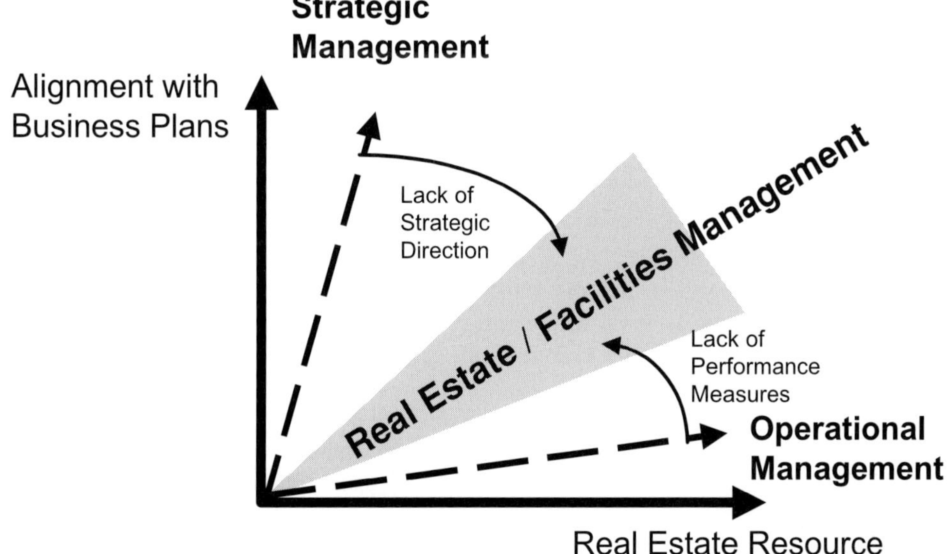

Fig. 5.2 Conceptual representation of the communication gap.

a lack of clear performance measures in relation to the supply and utilization of operational facilities, often results in situations in which the efforts of operational managers are not aligned to the objectives of the strategic plans. The following reflects these weaknesses.

- The difference in the perceptions of senior management (as clients), business units (as customers) and staff (as consumers of workspace) of the role of operational assets.
- The difference in the perceptions of these groups to real estate as an asset, i.e. the motivational drivers that influence the emphasis given to the management of operational property – viewed as a physical product, viewed as a business resource, or viewed as an organizational enabler.
- The different planning horizons imposed by strategic management and operational asset management.
- A mismatch between particular business operational requirements and the attributes of the existing real estate portfolio.
- Organizational influences on the delivery processes of operational property and facilities support services.

There is evidence from research (listed in Appendix A) carried out in North America, the United Kingdom and elsewhere that points to a growing awareness of the need to incorporate facilities thinking into the strategic business planning process. The promotion of this awareness emanates from the gradual awakening of business to the critical role that the workplace environment can play in advancing corporate teamwork and enhanced productivity, a subject we have discussed in other chapters. In our opinion, this shift in mindset owes more to competitive survival, advancements in the field of organizational development and human resource issues, than it does to factors relating to real estate. In particular, for many businesses the critical issues are those of the relationship between the retention of key staff and workplace quality, as well as those relating to the economics of premises occupancy governed by what is affordable rather than what is desired. The pace of change, increasing global competition and the intensification of turbulence in the local business environment, also demand a need for flexibility and adaptability in the supply and ongoing management of corporate facilities and support services.

Apart from the corporate perception of the role of operational assets, there are also the external dimensions of the market environment, which influence strategic business management. For any business, the arena of strategic management is the environment external to the organization, and the market forces that impact upon the products and services of the business. In terms of the impact on operational assets, the main concerns are those factors that result in a change in the size and nature of the existing building portfolio. The origins of such factors are a direct result of the business strategy that is developed in response to external market forces, and their strategic evaluation by the business. The output of this evaluation by senior management is commonly reflected in a shift in strategic direction of the business. The focus of such a shift can be viewed in terms of three possible outcomes or responding corporate objectives.

- **To ensure continued corporate survival** – the elimination of waste and surpluses, and the minimization of risk.

- **To remain competitive** – by responding to, and investing in, developments in technology and innovative management techniques.
- **To plan for growth** – expansionist and risk taking, investing in the future.

Each of these management initiatives will have different implications for the specific measures to be taken by the business to realign the nature and mix of the required supporting operational asset base. The specific measures taken, in terms of operational asset management, must demonstrate how they can respond to meeting the specified corporate objectives. It is noted that the *push* is very much uni-directional from strategic management. The context of this uni-directional push and its implications for operational asset management can be largely attributed to the separation between strategic management processes and operational management processes. Evidence supports the view that this separation is often reinforced by the perception, on the part of the senior management of the business, that building assets and facilities services have a non-strategic role within operational management. Making decisions at a strategic level, without integrating the property and operational dimensions of the business, clearly contributes to the development of sub-optimum solutions in many organizations. In so doing, this reduces the role of the real estate/facilities function to one of reacting to the demands of the business rather than helping to shape them. However, research and case studies (Then, 1996) suggest that for some organizations there is an emerging development where solutions to key management initiatives are increasingly reflecting, not only operational elements, but also the strategic importance of those operational assets and facilities services. Their importance is seen in their ability to support and contribute to business restructuring, and hence improving corporate effectiveness through improved utilization resulting from a more focused matching of supply to demand.

Figure 5.3 shows some key management initiatives emanating from corporate objectives and their implications for the supporting operational asset base. The specific measures under the control of facilities management must demonstrate how they can

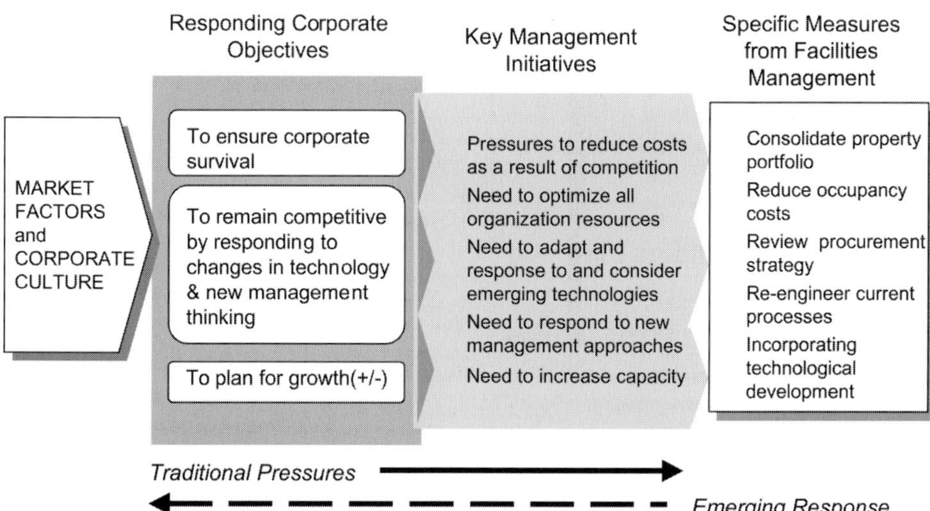

Fig. 5.3 Strategic response to management initiatives.

contribute to meeting the identified corporate objectives by saving costs, or adding value, and increasingly both. It is crucial to acknowledge that the specific measures identified must be directly related to achieving the key business direction intended by senior management. In this way, corporate objectives and the resulting key management initiatives will underpin any measures relating to the operational asset base. The feedback loop is closed when the outcomes of such measures are reported to the senior managers of the business, and are seen by them as contributing to the attainment of corporate objectives. It is in this context that Fig. 5.4 proposes an outcome that is the result of the integration of decisions relating to strategic choices, with these choices relating to the economics of facilities provision and management. The role of strategic facilities planning is central in bridging the much-needed dialogue between the corporate strategic choice and supporting strategies for the delivery of facilities and support services within the business – the outcome being well thought out and evaluated supporting facilities strategies that continuously meet the business requirements through fitness for purpose and flexibility in supply.

The challenge for facilities managers, architects and business managers alike, is how to deliver a workplace strategy that uses all available business resources to their fullest potential. Research suggests, and our experience confirms, that workplace solutions that are effective are founded on a thorough understanding of the work processes of business and the *real* needs of the customers, both internal and external. It is against this framework that we will consider the supply variables of space planning.

5.2 Location of workspace

It may at first seem strange that the premises audit includes consideration of the location of the building. As the building does not move, or at least we hope it does not, then surely it remains in the same location it has always been in. There are two key reasons for

Fig. 5.4. Emerging response to generating supporting facilities strategies.

reviewing the location of the building. First, while the building may remain static, its surrounding environment does not. Redevelopment of adjacent land and buildings, the introduction or the removal of transportation systems and new legislation, all change the environment in which the building is set. Secondly, the needs of the occupying business can change in terms of its markets, and access to labour and materials. These too will affect the locational criteria that should be applied to the business's workspace, since what was appropriate in the past may be inappropriate in the future.

Typically, the location of business premises should be assessed against the following minimum criteria.

- The proximity to pools of labour with the necessary skills.
- Services by both public and private transportation, in and out of the site; this should include an assessment of both staff and visitor parking as well as goods access and unloading facilities, and connections with national and international networks.
- The relationship to customers, competitors and other locations of the business.
- Services infrastructure, in particular public utilities and communication systems; the ease with which services can be delivered to the premises and upgraded in the future – for example, could the building meet the requirements for line of sight satellite communications or are there too many building obstructions in the way.
- Is the location suited to an easy re-sale or re-letting of the premises – what was once a desirable area can be adversely affected by surrounding development or the lack of it, and consequently, a good re-sale or re-letting location when the premises were first occupied, may not stand the test of time.

When considering the business drivers for location of corporate workspace, the facilities manager needs to give due consideration to the potential that exists for non-traditional business locations, which may now and in the future influence the selection of premises. By way of example, since the mid-1990s many businesses in the UK financial services industry have broken free of the conventions that would have guided them to locate in inner cities, in the same way as manufacturers would conventionally site their premises on industrial estates. For many, the change has been led by the development of call centres located either on the outskirts of cities or in provincial centres rather than in major urban conurbations, bringing a surge in the relative economic prosperity of these fortunate locations.

As businesses come to terms with the fact that the number of people they require at the corporate centre can reduce as well as increase, they can be faced with an experience that many may not have encountered before, namely a surplus of workspace. Where the excess can be consolidated into a whole floor, or on a multi-building site into a whole building, then the prospect for 'releasing' the space is enhanced. In many instances consolidation may not be practical, resulting in pockets of workspace being left vacant. Even where consolidation of surplus space can be achieved, its occupation by a third party may not be considered desirable in terms of the implications for security, and shared use of building services and utilities. Some businesses have, however, seen potential in another approach.

Case Note 5a

When Lloyd's, the operator of the insurance market in London, was faced with almost 2000 sqm of surplus workspace in their head office, they addressed the matter by a bold and imaginative initiative. Following an initial approach by Regus, the internationally renowned operator of high quality serviced workspace, the parties forged an innovative agreement. In June 1998, by which time Lloyd's had consolidated the surplus workspace into one homogenous area, Regus commenced a 10 year lease of the workspace, during which time they will market and operate the space as a 'business centre'. The fully serviced workspace is available for hire by third party businesses for short-term periods ranging from a few days to a few years as their needs demand. Regus manages the serviced space providing a full range of facilities for their tenants, some of which are provided by Lloyd's in-house facilities team and which are charged to Regus on full commercial terms (see Fig. 5.5). The benefits of this arrangement are manifold. For Lloyd's, apart from the elimination of a non-performing asset from their estate, as Nick Phillips of Lloyd's puts it, 'Regus provide a well known branded product to which buyers of serviced workspace are accustomed. This it is hoped, will result in occupiers being attracted to the premises who may eventually seek long-term accommodation which Lloyd's is well placed to provide from their principal facilities.'

For Regus, they have a high quality building with a sophisticated infrastructure in a prime location, in which to develop their flexible workspace offering for the business market. For tenants, they are able to operate from first class accommodation in the heart of the business district of London, equipped and serviced to a high

Fig. 5.5 Lloyd's building at One Lime Street, London, incorporates a business centre managed by Regus, which Lloyd's, as well as third party businesses can use.

standard, and all on flexible 'pay as you go' terms. Of course such arrangements are not without risks. To minimize the more obvious ones, access to the serviced suite is via an entrance that is separate from that to Lloyd's own space, and tenants are not permitted to refer to Lloyd's in their address.

Arrangements such as the Lloyd's–Regus one will, we believe, become more commonplace as businesses come to appreciate the mutual commercial benefits that can be derived from sharing workspace in moves to reduce the fixed cost elements in their business. In such a world, shared business space, be it serviced offices, serviced call centres or serviced factories, will be an accepted feature of facilities management.

5.2.1 Location attributes

The historic trend has been for workspaces of similar types to be congregated together because of the business interactions and synergies that existed. In the past, the creation of 'vegetable street' or 'tin street' where purveyors of the same goods or services were co-located, which can still be seen in some middle eastern and far eastern cities, was born out of necessity. This has been further fuelled in the second half of the 20th century by the actions of town planners who have consciously structured cities and towns into zones of demarcation such as office, light industrial and manufacturing. In many western cities the same applies, where banks, lawyers, newspapers and many other sectors of business have grouped themselves together within the same 'quarters' of cities – e.g. the financial districts of New York, Frankfurt, Tokyo and the City of London. A key feature was the ease of exchange and interaction stemming from the days when so much of trade and commerce was conducted on a direct person to person basis, i.e. face-to-face.

This is no longer the case, or at least it need not be so, as technology and – particularly communications – has rendered much of the physical transaction and direct personal contact unnecessary. Nonetheless, some businesses for cultural reasons still seem wedded to their historical roots. However, the combined pressures of environmentalism and the search for increasing productivity are influencing many organizations to question the merits of remaining in such locations with all their inherent problems of traffic congestion, and pollution, let alone the current high accommodation costs associated with many such locations. We say current, because price is a function of the supply and demand cycle, and as businesses seek alternatives to these 'historic' locations in which to base themselves, then we expect to see direct accommodation costs, i.e. rents and local taxes start to decline. The magnetism of business synergies is well known. It can 'attract' organizations together, such as in the UK insurance industry, where broking and underwriting businesses 'cluster' around Lloyd's located in Lime Street in the City of London, where the underwriting market is based. Many businesses would cite close proximity to the Lloyd's building (defined in terms of five minutes walking distance) as being a primary deciding factor in their location of business premises. And so no doubt many facilities managers would diligently include this within their data capture process for business premises criteria. However, how many of these businesses and their facilities managers have assessed what they would do if Lloyd's relocated their premises? Or if the underwriting market itself decided to move premises?

So when considering the business drivers for location of workspace, the facilities manager needs to give due consideration to the external 'magnets' that could influence co-location of their premises, as well as the potential that exists for non-traditional business locations which may now, and in the future, influence the selection of premises. Therefore, when assessing the needs of the business in terms of premises location, the facilities manager needs to look for signs that could point to such occurrences or alternatively could impede and make such an approach inappropriate.

Transportation is a major factor influencing workspace locational decisions. The ease of getting workers to the workplace, as well as products to customers, is of paramount importance to all businesses. When early occupiers moved to premises in London's dockland redevelopment areas, they soon became aware of the constraining effect that inadequate transportation systems can have upon business. Difficulties in recruiting staff, and then retaining them, poor access for customers and suppliers, all stemming from inadequate road and rail links, impacted upon the operations of these 'early settlers'.

The internationalization of work and work processes is all around us, as product components are manufactured in one or more countries and shipped to another country for assembly. Japanese car manufacturers have used this technique to gain a very sizeable share of the North American and European markets, where they not only find ready buyers for their products, but are also able to take advantage of lower labour costs. The globalization of work means that the workplace can be located anywhere within a single country, or even worldwide, as the historical convention of aggregating more and more people in one location is being challenged by the concept of dispersal, for reasons which include the following.

- Environmental pressures to reduce traffic congestion with the ensuing consumption of energy and resulting pollution.
- Personal pressures to reduce travel time in favour of an increase in productive time and quality personal time, at home or wherever.
- Lower unit costs of production, for a long time the watchword of manufacturing, is also now a feature of service and administrative 'office'-based businesses.

In the past, location of business premises has been generally considered in terms of at least three dimensions.

- Convenience for business transactions and markets.
- Convenience and proximity to appropriate skills, labour, raw materials and the means of transportation.
- The cost of provision in terms of the conditions prevailing in the property market and the management of occupancy costs.

Deliberation of these factors appears to have altered the relative weightings and, indeed, challenged the conventional wisdom that has influenced the location of business premises. Technologies, like fibre-optic communications, powerful and portable personal computers and cellular telephones, have opened up a variety of possibilities for accomplishing work and given us products that have, in turn, led us to question business's dependency on traditional locational attributes. The electronic highway of E-commerce offers completely new ways of doing business that disregard the constraints imposed by local, regional, national and, increasingly, international boundaries.

5.2.2 Impact of changing work patterns

Most workspace today has been designed and is managed, on the basis of a set of assumptions that have been the subject of radical change over the last decade. Today, organizations seem to be changing their form faster than ever, in order to compete in a rapidly evolving global economy. In response to the vying demands for reduced occupancy costs, improved productivity and increased worker satisfaction, many organizations are experimenting with new space strategies. It is the facilities manager who is often made responsible for the transformation from the past to the future. Within the corporate real estate and facilities management professions, it is becoming increasingly commonplace to hear discussions of alternative ways of working, flexible workforces and the teleworker. More and more space strategies have been evolved by organizations to cater for new demands arising from the new ways of working, particularly for office workers.

The term 'alternative officing' has been used as a broad concept encompassing a range of innovative approaches that focus on how, when and where people work. The emerging space strategies are fundamentally in response to the many ways in which people today relate to the organizations that employ them. Management hierarchies are delayering, with fewer job grades; for many, the variations in 'flexible' employment contracts, largely as a consequence of outsourced services, have changed the pattern of the standard working week and year. The emphasis on the time dimension in addition to space (the area dimension), has challenged the long accepted concept of providing dedicated space in an office building for particular individuals. Table 5.1 (HOK Consulting, 1993), offers a convenient summary of the pros and cons of alternative officing strategies grouped under the headings:

- Re-engineered space;
- On-premises options;
- Off-premises options.

It is clear that one size does not fit all. For some organizations, off-premises options are not yet feasible alternatives to the traditional work environment because of the prevailing corporate culture. For others, the sales force may already be working in a virtual office mode and the question is, which options will work for the rest of the organization. In considering the relevance of both cases, and all other variations in between, it is important to appreciate that each organization should be treated separately and analysed individually to determine the most feasible solution for implementation with a successful outcome. This also illustrates the significance of the time dimension as a variable in the optimum supply of workspace, which is necessary to achieve the fulfilment of a variety of tasks within different work settings at numerous locations. The most significant factor in influencing the changing geography of workplaces, has been the impact of new technologies. Their development continues to shape the nature of work and have profound implications for the buildings that accommodate and support business processes. So how well do these buildings meet the demands of business? And how can you identify the ones that do support business operations, from those that do not?

5.3 Building attributes

The attributes of all buildings are shaped by their design and influenced by the manner in which they are maintained and serviced and, as a result, their prevailing condition.

Table 5.1 Alternative office strategies

Type of strategy	Organizational and individual advantages and benefits	Organizational and individual disadvantages and pitfalls
RE-ENGINEERING SPACE		
Work schedule Flexible work schedules and timing	• Take advantage of additional segment of labour interested in part-time work • Work is accomplished when worker is most productive • Reduction in transportation costs • Improve staff's quality of life	• Problems may arise that require attention when staff is not in the office • Requires staff to structure time and work differently from past
Modified workplace standards Refined standards to improve productivity and efficiency	• Reduced space required with better utilization • Less hierarchical space allocation • Increased flexibility for staff moves • Decreased move costs	• Organization cannot use workplace size to indicate seniority • Inhibits customization of workplace
ON-PREMISES OPTIONS		
Shared space Two or more staff sharing a single, assigned workplace	• Improved utilization of workspace • Increased headcount without increasing required workspace	• Staff may be reluctant to give up 'own' workplace • Requires staff to work closely with one another, maintain files and workspace in orderly manner
Group address Designated group or team space for specified period of time	• Project team orientation of users ensures high ongoing utilization • Encourages interaction between team members	• Size of teams may create workspace shortage • Difficult to manage turnover of workplaces among users • Requires accurate projections of user group size and volume
Activity settings Variety of work settings to fit diverse individual or group activities	• Provides users with a choice of setting that best responds to tasks • Fosters team interaction	• Requires advanced technology • File retrieval can be problematic
Free address Workplace shared on a first-come first served basis	• Maximizes use of unassigned workspace • Minimizes real estate overhead • Minimizes cost of workplaces and office construction • Suitable for sales and consultant practices	• Access to files and storage can be problematic • Probability high for scheduling conflict • Substantial investment required in equipment and training
Hoteling Reserved workplace	• Accommodates staff increases without corresponding increase in facilities and property costs • Can result in upgrade in office amenities	• Storage can be problematic • Probability high for scheduling conflict
OFF-PREMISES OPTIONS		
Satellite office Office centres used full-time by staff closest to them	• Lower rentable costs / sqm • Reduces communication time	• Remote management of staff can be a challenge • Staff may feel disconnected from organization • May impede ease of inter-office communication
Telecommuting Combination of home-based and office workplaces	• Transportation and real estate costs reduced • Improved quality of personal life • Potential increase in productivity	• Inadequacy of home environment • Staff interaction is reduced • Home workspace must be quiet and free from interruption
Remote telecentre Office centre located away from main office, closer to customers; workers can access technology and support	• Reduces transport pollution • Fosters productivity and staff loyalty through improved family life	• Clear guidelines for and support of supervisory staff are required
Virtual office Freedom to work anywhere, any time supported by technology	• Increase staff productivity • Potential increase of time spent with customers due to reduced commute time • Reduced workspace and attendant costs	• Staff may feel disconnected from organization • New criteria required for evaluating individual performance • Reduces staff interaction

Consequently, a condition survey should form part of every premises audit, the extent being governed by what is identified in an initial review. The principal issues to be addressed for all building elements will cover the following.

- **Wear and tear** – have damaged elements been repaired in time, or is there evidence of advanced deterioration resulting from failure to make adequate repairs in the past.
- **Maintenance** – this should not be confused with repair works, since maintenance is necessary for all assets to ensure they are operating at peak performance; failure to maintain building elements adequately can lead to an acceleration in their loss of performance and give rise to breakdown and critical failure, which could have been avoided if adequate maintenance had been applied.
- **Replacement** – in this context, replacement means exchanging at the end of the natural life of the building element, as opposed to replacement due to damage or inadequate maintenance; this will involve identifying the current position in the life-cycles of each of the building's key components.

The premises audit will assist in the preparation of planned maintenance and investment plans, as it will identify early signs of decay and give advance warning of elements reaching the end of their natural life. The condition survey should aim to identify the interrelationship of the three principal elements above, which although considered here in isolation inevitably interact one with another. The outcome of the survey should be a report detailing the current condition of the premises, inside and out, and the courses of action to be pursued to ensure the premises remain as assets to the business.

The accelerated move towards diversification of workplaces will clearly place greater demands on building performance than in the past. The quality of a building and the functionality that it possesses is at least as important as the amount of space it provides. When considering the suitability of buildings to meet the needs of the business, there are three key elements that should form part of an appraisal.

- Building value.
- Technical suitability.
- Functional suitability.

5.3.1 Building value

Understanding the role of buildings and how they can be deployed effectively is the essence of what facilities management is about, in the context of the operations of each individual business. It is essential for businesses to realize that buildings are necessary not for their own sake, but as *supporting resources* through the provision of workspaces to accommodate business processes. And in so doing, to provide appropriate workplaces at which the people in the business can perform their tasks to the best of their abilities. This may seem self-evident, but the evidence of many businesses' work environments would suggest otherwise. Figure 5.6 illustrates the positioning of buildings in the context of business (Then, 1994).

As a supporting corporate asset, whether owned or leased, buildings attract liabilities if not properly managed. Prudent asset management – for buildings are assets – is therefore a requisite in order to ensure optimum utilization of the resource, to sustain its continued

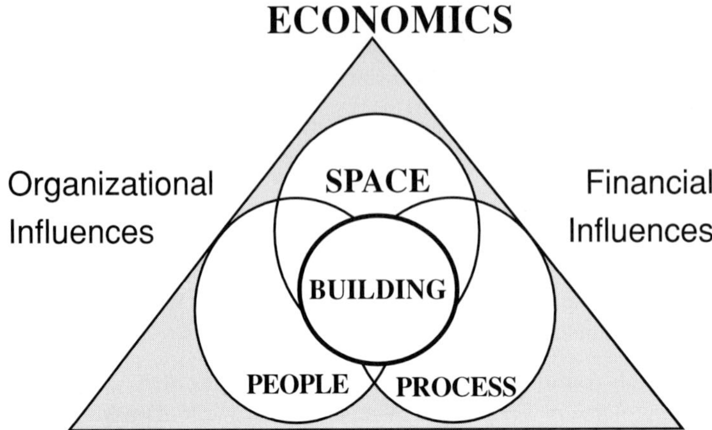

Fig. 5.6 The positioning of buildings in business.

functional suitability, and to protect or even enhance the asset's intrinsic worth. In this respect, the economics of operational asset management can be viewed from several narrow perspectives: in terms of its exchange (financial) value, its operational (cost) value, and its organizational (utilization) value. In practice, the true economic value of a built asset must be seen as a result of composite measures that reflect its contribution to the value of the business, comprising a combination of all three components of value. Buildings must therefore be managed as value-adding facilities, and not as consumers of vital resources.

The durability of buildings and their locational fixity may conjure a notion of inflexibility. In reality, buildings are creatures of time, and thus of change. As a dynamic business resource, the focus on buildings in-use requires ongoing management, continually adapting to provide affordable and appropriate services to their owners and users, whose needs change as economic conditions change. Figure 5.6 illustrates the scope for building appraisals in meeting the dynamics of the work environments in which business must operate. The performance measures chosen, must reflect a balance between supply and demand, and between variables that denote the *hardware* elements of buildings (as a physical product) as well as the *software* elements of user requirements (as an enabling working environment).

5.3.2 Technical suitability

In viewing operational buildings as a business resource that can be used by organizations and individuals to achieve their goals, it is important not to regard the product as being static; that is, unchanging. As a physical product, it is unique in terms of location, design features and relatively high cost of creation. As an economic resource, its potential benefits and liabilities must be considered in terms of financial viability over time. In this respect, the principle of life cycle asset management offers a valuable tool for measuring the economic worth of buildings and the options for investment in them over their operational life. While detailed calculations of life-cycle costing (Flanagan *et al.*, 1989) will assist in evaluation of investment options at building components level, such as the type of flooring material to use, the methodology becomes cumbersome and complex when applied to a whole building. Duffy (1990) provides a useful perspective of the principle of building life cycle management, by considering a new building as comprising four groups of layered elements (shell, services, scenery, settings), each with their own typical life cycle, to which we add a fifth, the site.

- **Site** – the external surroundings to the building that set its environmental context, and from which the macro infrastructure of utilities and transportation are provided.
- **Building shell** – the permanent structure and enclosure of the building which lasts 50–70 years, although many components that comprise the building shell may have shorter lives, 20 years or so.
- **Services** – the heating, ventilation and cable infrastructure of a building, which may have a life span of 15 years or less, before the technology becomes redundant.
- **Scenery** – the fitting-out components, which adapt a building shell to the specific requirements of an organization, with life spans of 6–7 years.
- **Settings** – the day-to-day timetabling, management and rearrangement of furniture and equipment to meet specific work processes and tasks.

Figure 5.7 illustrates the layering of the five elements comprising site, shell, services, scenery and work settings and their typical life spans. Analysed over time, in a new building, costs can be roughly divided into thirds – one third for the 50-year shell, one third for the 15-year services and one third for the 5-year scenery. The costs associated with changes to the work settings (re-arrangement of furniture layout and IT infrastructure) which may occur daily, are now commonly regarded as the *cost of churn* within organizations, a subject which we will be discuss in Chapter 7. Importantly, when the cost profile of the groups of elements are appraised over the invested life span of a building of say, a 50-year period, the shell expenditure is overwhelmed by the cumulative financial consequences of three generations of services and ten generations of scenery as illustrated in Fig. 5.8. (Duffy, 1990).

As Duffy (1990, p.18) aptly points out, 'It is this same iron economic logic that explains why large areas of the City of London are being demolished. 20-year-old buildings are being torn down despite their apparent permanence because, in terms of use, they are judged to be prematurely obsolete, and because it is more important to get the services and scenery right than to preserve the financially less significant shell.' It is of paramount importance for facilities managers to grasp the significance of this statement. Understanding the building life cycle profiles, in terms of shell, services, scenery and settings, is vital not only in terms of cash-flow and outlays in asset management terms, but to recognize that the real prize is about creating workplaces in which people can perform their tasks effectively and interact productively.

5.3.3 Functional suitability

In the book, *Reinventing The Workplace*, writing on the subject of 'New patterns of work: the design of the office,' Andrew Laing (1997) describes the 1990s as the period when the tyranny of supply-driven developments, which dominated the UK and North America

Fig. 5.7 Site, shell, services, scenery and settings.

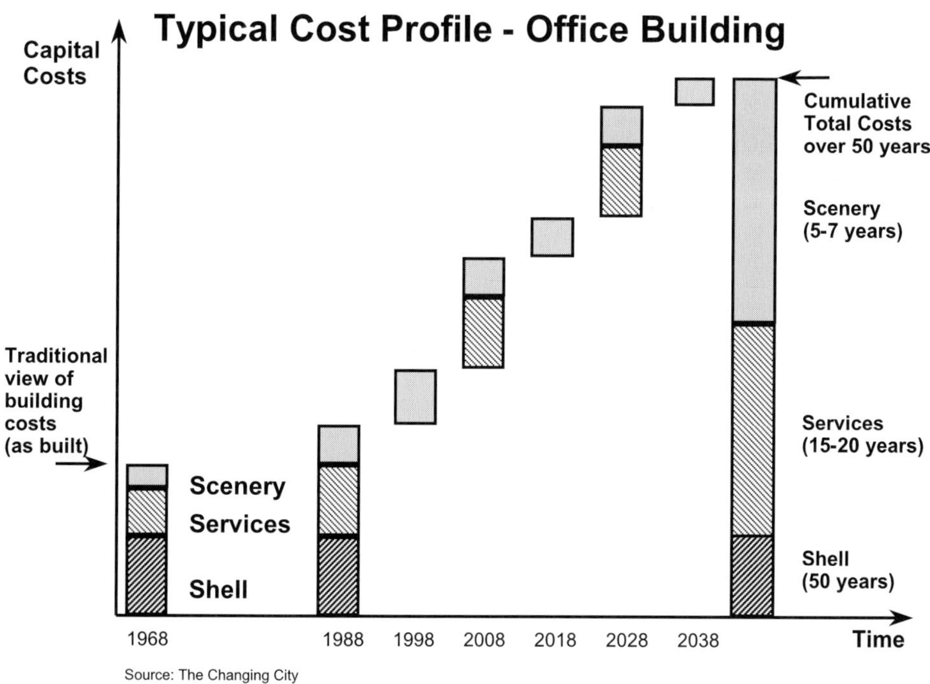

Fig. 5.8 Cumulative expenditure on shell, services and scenery.

throughout the 1980s, was broken. Office design is at a pivotal point in its history, as a new world of work is propagated in which organizations can choose to manage work across time zones, on multiple sites, and in a variety of settings.

In Chapter 2, we traced the development of the workplace from its beginnings as essentially a paper-factory, to the present-day rapid growth of call centres. In the intervening half century, technology has changed the way businesses are organized; yet in their physical form work buildings have changed little. In speculative office developments the form of construction has tended to favour steel-framed structures with predominantly glass claddings. Over the same 50-year period, developments in the environmental systems of lighting, heating and air-conditioning, and the ergonomics of furniture design, have perhaps progressed more so, benefiting from investments in research and development.

Given that the essence of space planning is the fitting of an organizational structure into a building structure, then when choosing buildings to support their customers, apart from locational attributes, the facilities manager must also appraise the building shell, the building shape and the layout of its floor-plate. These building attributes are key determinants that objective evaluation can identify as the opportunities and constraints of individual buildings to meet organizational, user and statutory requirements. In space planning terms, the building appraisal, sometimes called the audit, can reveal how space within the premises can be configured to provide appropriate workplace solutions for the organization. In a broad summary of building types in North America, UK and North Europe, Duffy *et al.* (1993) observed that buildings that have been shaped by direct user influence in Northern Europe are significantly different from those of the supply-biased, developer-oriented offices of the USA and the UK (Fig. 5.9).

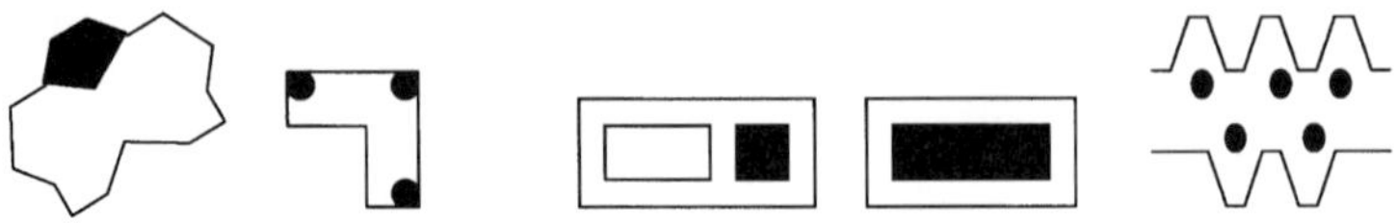

	Bürolanschaft offices	Traditional British speculative offices	New 'Broadgate' type of British speculative office	Traditional North American speculative office	The new North European office
No. of storeys	5	10	10	80	5
Typical floor size	2000 sqm	1000 sqm	3000 sqm	3000 sqm	Multiples of 200 sqm
Typical office depth	40 m	13.5 m	18 m and 12 m	18 m	10 m
Furthest distance from perimeter aspect	20 m	7 m	9–12 m	18 m	10 m
Efficiency: net to gross		80%	85%	90%	70% (lots of public circulation)
Maximum cellularisation (% of usable)	20%	70%	40%	20%	80%
Type of core	Semi-dispersed	Semi-dispersed	Concentrated: extremely compact	Concentrated: extremely compact	Dispersed: stairs more prominent than lifts
Type of HVAC services	Centralised	Minimal	Floor to floor	Centralized	Decentralized: minimal use of HVAC

Fig. 5.9 Planning and design criteria for various types of office (Duffy *et al.*, 1993, p. 9).

Table 5.2 lists some of the considerations that should be taken into account when appraising a building's physical features in terms of its building shell, its shape and its floor-plate.

Table 5.2 Building attributes of shell, shape and floor-plate

Shell	Shape	Floor-plate
• Type of structure • Type of cladding (skin) • Construction and cladding materials • Floor-to-ceiling height • Number of storeys • Structural grid • Planning grid • Appearance	• Site constraints – access, car parks, security, aspect • Shape of building • Layout options • Ease of sub-division • Space efficiency net:gross • Net Usable Area	• Shape of floor-plates • Size of floor-plates • Location of stair and service cores • Building depth • Sizes of contiguous elements of space

Case Note 5b

Financial services company, London. In 1995 this financial services company relocated to new head office premises, which will for all time add a burden to their facilities operating costs. The reason for this penalty is that the building demonstrates poor provision of workspace, in that 6% of the net area of the building is unusable, resulting from wasted space in the form of gaps (some less than 800 mm deep) between the perimeter columns of the building and the internal face of the external wall, considered a 'design feature' by the building's creators. This space is of no productive value to the occupying business as they are unable to make any use of it, but they still have to maintain and service it. The estimated rent for the space throughout their tenancy of the building over a 25-year lease period could amount to over £5 million, to which would be added the even heavier burden of facilities running costs. Clearly if the occupier had conducted a thorough appraisal of the building prior to committing to the lease, they could have identified the impact the wasted space would have upon their long-term operating costs. However, this example also points to a failing in some parts of the property industry, to appreciate the consequences of their actions. Providing buildings of 'funny shapes' which result in wasted resources (financial and environmental) is not in the interests of the industry or of its customers, let alone of society at large.

5.3.4 Environmental 'green' issues

Increasingly it is being realized by both the property industry, i.e. the providers of buildings, as well as the occupiers of buildings, that workspace needs to be provided in an environmentally responsible way. In addition to measures that foster energy conservation, such as appropriate levels of thermal insulation and efficient space heating/cooling systems (both passive and active), the producers of buildings can also make significant contributions to improving the environmental impact of their creations through adopting best practice in the use of materials. The adoption of schemes such as the Eco-Management and Audit Scheme (EMAS) can make valuable contributions when incorporated into the design and development process. By the specification of materials that carry the EMAS logo, producers of buildings can support bio-diversity through the use of naturally sustainable materials and actively help to reduce 'product miles' by sourcing locally available materials. Such measures help contribute in very practical ways to local and global environmental improvements resulting from the minimization of waste, lost energy and lessening the pressures to find suitable landfill sites. 'Eco-awareness' should not just be seen as a feature of the development of a new building, but should feature throughout the life of a building, i.e. from its inception to its construction, and then throughout its lifetime operation to its eventual demise. The concept of responsibility for lifetime environmental impact is at the very core of BREEAM, the UK's Building Research Establishment's Environmental Assessment Method. This mould-breaking scheme, first introduced in 1990, sought to guide the design and development of office buildings in ways of minimizing the adverse effects that they had upon the local and global environments. The scheme, a world first, was subsequently extended to encompass other building types, including industrial units, supermarkets and domestic dwellings, and has been emulated in countries as far apart as Canada and Hong Kong.

Initially, the BREEAM scheme sought to assess proposed buildings (new and refurbished) on key criteria, the measurement against which would guide building owners and designers on the environmental effectiveness of the building, through a scoring of its design and operating parameters. The principal aspects, assessed under the three headings of Global, Local and Indoor issues, included the following.

- Minimization of carbon dioxide emissions resulting from energy use.
- Minimization of ozone depletion and acid rain.
- Use of natural and sustainable resources and recycled materials.
- Designing for longevity.
- Minimization of overshadowing of adjacent buildings and land.
- Minimization of the impact upon the ecology of the site.
- Water conservation.
- Minimization of noise.
- Air quality and ventilation.
- Avoidance of the use of hazardous materials.
- Appropriate balance in the use of natural and artificial lighting.
- Provision of thermal comfort and the avoidance of overheating.

While all of these factors remain laudable goals to which businesses should aspire in the provision of their workspace, the BRE has developed the scheme in a move towards what it calls 'full life' assessment. In the light of experience gained from operating BREEAM, its further development has moved away from the Global, Local and Indoor categories of assessment, to a set of criteria that the BRE considers relate better to the concerns of the property industry and to environmental impacts. The existing BREEAM schemes, commencing with office buildings, have been changed to incorporate a core assessment of the building fabric and services with two further optional parts relating to (a) the design/procurement of the building and (b) the management/operating procedures employed. 'BREEAM 98 for Offices' therefore establishes a set of categories under which specific credit requirements of the rating system are grouped.

- **Management** – Overall policy and procedural issues.
- **Health and comfort** – Indoor and external issues.
- **Energy** – Operational energy and CO_2 issues.
- **Transport** – Transport-related CO_2 and locational issues.
- **Water** – Consumption and leakage-related issues.
- **Materials** – Environmental implications of materials selection.
- **Land use** – Greenfield and brownfield site issues.
- **Site Ecology** – Issues relating to the ecological value of the site.
- **Pollution** – Air and water pollution issues (excluding CO_2).

In each category there are several credit requirements that reflect different options available to designers and managers of buildings. Buildings submitted for rating under the scheme are evaluated under each heading, leading to an overall BREEAM rating for the building. This approach therefore enables the objective comparison of a composite design of a building with another of a similar type.

As William McKee, Director General of the British Property Federation says (Edwards, 1998)

> Green buildings will become commonplace only if they take account of the principles and philosophy of the commercial property industry. Sustainable development is unlikely to materialise if it ignores the economic constraints under which the industry operates. In particular, the decision to provide new commercial buildings is judged against their performance as an asset and their ability to attract tenants, thereby generating good rental levels and ultimately, adequate yields.

In short, green buildings have to *work* for their users. Property developers and landlords alike have not always appreciated this. For too long the property industry has been plagued by the age-old 'landlord versus tenant' debate, the nub of which focuses upon the balance between the initial capital cost of a development and its long-term running costs. The issue being that landlord/developers have historically ignored the potential benefits that may accrue to building occupiers through prudent investment in the initial infrastructure and services of a building, and have instead pursued a course based upon least cost of construction. Not only is this approach often environmentally irresponsible, but it does not make good business sense. Many building occupiers, if given the chance, would be prepared to pay an increment on their rent and/or service charge in the knowledge that they are being provided with a more efficient, cost effective and flexible building infrastructure. Alan Rowe of Landsdown Estates, a subsidiary of MEPC, the UK's largest property investment company, advises 'Capital costs, running costs and depreciation costs mean that Landsdown Estates look to eliminate the requirements for air-conditioning [in buildings] where possible. However, we do design our buildings so that they have an upgrade path and can adapt to full or partial air-conditioning should occupiers so require.' As this approach to property development has so much to commend it, we question why more players in the industry have not seen fit to adopt it and extend it into other spheres by providing migration paths for upgrading IT, lighting and other services.

In '*Green Buildings Pay*', Brian Edwards (1998) considers that investment in green buildings can pay off in the first 8–10 years of the building's life, through resultant reductions in running costs. Clearly there exists a very wide range of energy consumption experienced by businesses in operating their buildings. While typical UK office buildings consume around 200 kWh/sqm over a typical year, best practice, as typified by the BRE's new premises (The Environmental Building, Garston, UK), can achieve levels of 80 kWh/sqm, with a realistic benchmark being 100 kWh/sqm (see Figs. 5.10 and 5.11).

Benefits accruing to businesses from their occupation of green buildings do not stop at their running costs. Buildings designed on environmentally responsible principles are, according to Edwards, healthier to use, provide the occupants with the psychological benefit of 'feeling better', while enhancing the corporate image of the occupying business. This view is also supported by the UK's Department of the Environment, Transport and the Regions (DETR, 1998) in their listing of the key benefits of naturally ventilated buildings.

- Reduced capital costs.
- Reduced operating costs.
- Reduced environmental impact.
- Creating the opportunity for occupant control leading to enhanced satisfaction and productivity.

However, we consider that potentially the greatest benefit of all is the sustainability that can be achieved through the design and development of buildings that possess capabilities to enable them to be adapted and re-used for different purposes, and in so doing extend their effective lives.

In assessing the capabilities of buildings – the supply side of the workspace equation – businesses should avail themselves of schemes such as BREEAM. We believe that one day all workspace buildings will be rated not just in terms of their environmental impact but also in terms of their fitness for purpose and future adaptability. This is not a new concept.

Fig. 5.10 The Environmental Building, for the BRE at Garston, demonstrates environmental responsibility in every aspect of its design.

Fig. 5.11 Readily controllable lighting and natural ventilation formed a key part of the design of The Environmental Building, for the BRE at Garston.

In the 1970s Alex Gordon, a past President of the RIBA, launched a study into a concept called 'Long life, loose fit, low energy' which was aimed at developing approaches to the design of sustainable buildings. As Gordon (1974, p. 9) puts it,

> ecological thinking is not limited to energy alone: it should involve being aware of the far-reaching implications of our current actions. Our predecessors left us with a stock of buildings which, in general, have been pretty adaptable and have served us for a long time. One suspects, however, that many new buildings will not be suitable for the functions for which they were designed for more than a comparatively short time.

Twenty years on, his words are both an apt testament to much of the workspace constructed in the intervening period, as well as applying to many buildings of the present generation.

Case Note 5c

BRE, Watford, England. In the development of their new premises on a brownfield site at their Garston campus, the BRE, as one might expect, demonstrate several examples of environmental best practice. Materials recovered from the demolition of the buildings that previously occupied the site were recycled. Some materials, such as crushed brick and concrete, were re-used as hardcore in the construction of the new building, while other materials were given 'new lives' elsewhere, for example old roof timbers were sold to a company that makes pine furniture and old electrical fittings were given to local schools for use as teaching aids.

5.4 Building utilization

Utilization is about achieving the optimum match between timing of need and availability of supply. In terms of space planning and management, the objective is to aim for the best match between demand for workspace to support business activities and its availability in terms of timing and duration of requirements. Demand for workspace relates to appropriateness of the types of space and their distribution within a building or a portfolio of buildings. Supply relates to the pattern of use of the available space and the range of tasks it is supporting.

Utilization is not only a function of efficiency of layout, but is also derived from the degree of flexibility the premises afford the users in terms of capability to support the relocation of working groups, departments and functions, and also to meet changing patterns and practices of work.

Some buildings afford little or no flexibility in cellular division since they impose too rigid a constraint upon the positioning of partitions. However, the flexibility afforded by appropriate planning grids does not in itself guarantee effective utilization of workspace. The planning grid module must be a function of the space standard to be allocated, if wasteful space is to be avoided. For example, if the space standard required for a given function amounts to 15 sqm but the planning grid only permits cellular space of 18 sqm to be created, then 3 sqm of space is wasted each time this workplace is created, leading to poor utilization of space. Similarly if flexibility within services (both environmental and information) is limited, then scope for adapting the premises to suit different working practices and layouts will be constrained. Assessing the suitability of premises to meet current and future work demands is therefore a key part of the audit process, the results of which will indicate whether the premises fit the requirements of the business or, where there is a misfit, whether they can be suitably re-engineered. Where insufficient flexibility and scope for adaptability exist, then the audit is likely to indicate that relocation and disposal of the existing premises may be the most cost-effective way forward.

The audit will also examine how efficiently space is used. Development of corporately endorsed space standards for each type of activity is of invaluable assistance in this regard since they can form a foundation not only for the allocation of space but also as a yardstick for measurement of utilization. However, the utilization factor also possesses a time dimension, i.e. for what durations are the building, and the workplaces contained therein, occupied? There are 168 hours in every week and that is the time that the real estate that is owned or leased by the business is available. Yet for most organizations it is probably only open for use for 50 hours a week, and then typically only occupied for 40 hours a week at best. To put it another way – most workspace is at best utilized for 25% of the time and that is assuming that staff are always at their workplace. This is a subject we will revisit in subsequent chapters.

In the 1990s a striking paradox emerged – workspace became more, and at the same time less, important to business. More important, as the fixed costs of property became 'millstones' round the necks of constantly evolving organizations. Less important, as technology enabled many elements of work to be conducted almost anywhere. In the context of the growing importance of buildings as a resource for business, the use of performance indicators to measure building effectiveness and utilization has also grown in sophistication. For many organizations the initial reliance on cost per unit area has given way to a family of measures that place emphasis on getting the best out of people at work – the most expensive resource in most businesses. Apart from the common average measure – the cost per person (i.e. the cost of accommodating an individual employee) – the challenge for facilities managers is to develop performance indicators that directly relate the output of the business to facilities inputs. Many manufacturing businesses use such measurements, where the cost of cleaning or catering, let alone the area occupied and hence the cost of property, is assessed in terms of its contribution to the cost of each product produced. Although widely practised in manufacturing, and by many retailers (sales per square metre of shop and warehousing space), other businesses have been slow to recognize the benefit of applying such indices. An example, in the health service, could be *facilities costs per critical operation*. In the education service, *facilities cost per pupil per year* may be a relevant measure. In an insurance company, *workspace area per person per policy transaction* may be appropriate. The development and use of facilities-related performance indicators also has the advantage of promoting the practice of 'looking outside of the box' by encouraging benchmarking and comparative analysis across organizations as a first step towards a culture of continuous improvement. However, any attempts at improvement must start with the careful identification and analysis of key variables that affect or influence the utilization of workspace.

5.4.1 Variables influencing space utilization

No matter how well a workplace is engineered to meet the needs of its user, it adds no value to the business when not in use. Experience from many office workplace utilization surveys shows that, for activities that are 'desk bound', occupants rarely spend more than 80% of their time actually at their workplace, i.e. a desk. Typically the balance of their time is spent accessing storage, using shared equipment, attending meetings (both on and off site), and having work breaks. Therefore, when allowance is made for holidays, training and sick leave, it can be said that the realistic *maximum* workplace occupancy that can ever be achieved is 80% to 85% of the average working day (see Fig. 5.12). Hence, when assessing workplace utilization, 'full utilization' must be taken as being in the aforementioned range and not 100%.

Nevertheless, in practice we encounter many businesses where the utilization of their workplaces falls between 40% and 50%. In other words, on average, half, sometimes more than half, of the business's workplaces are lying empty at any point in time. Such situations have led some businesses to adopt what are called flexible styles of working or new ways of working.

Initiatives aimed at improving space utilization can be considered under two broad categories. First, those that relate to changing organizational culture in respect of existing corporate conventions and ways of working. Second, those that relate to changing operational practices and procedures. Typically, the variables that impact on the first category of initiatives may include the following.

- **Corporate culture** – image, hierarchical organization structure, perception of the role of buildings and preferred configurations – space with status.
- **Occupancy level** – number of staff, range of tasks supported.
- **Intensity of use** – hours of operation per day.

Variables that impact on the second category of initiatives, i.e. aimed at operational practice and procedures, may include the following.

- **Ways of working** – functionally structured, team-based, project or process based.
- **Pattern of occupancy and tasks** – continuous or intermittent operation, the nature of the tasks and activities to be supported.

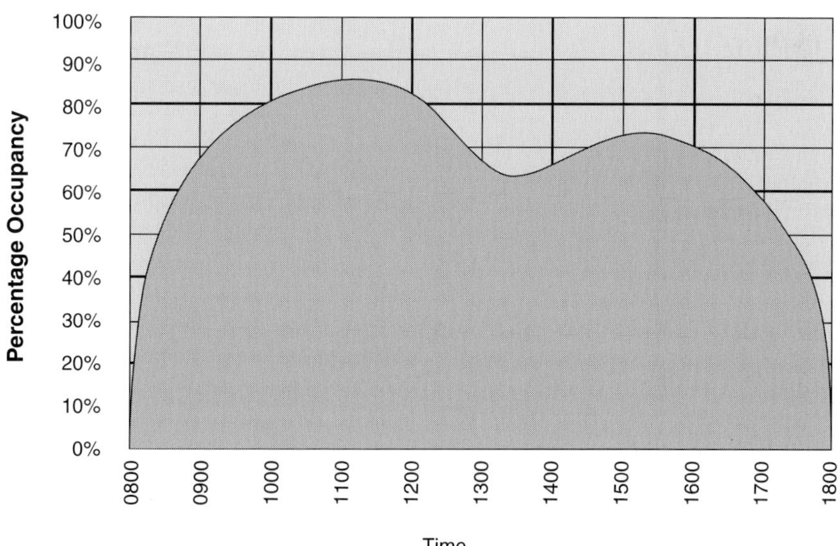

Fig. 5.12 The maximum utilization profile of conventional office workspace.

- **Appropriateness of tools** – access, layout and configuration, selection of products and sophistication of technology.

In reviewing the above variables it is crucial for facilities managers to understand the two underlying principles that give rise to the need for change. First there is the need to look at the big picture and take bold strategic decisions that reinforced the value of the physical resource for business success. Secondly, most sustainable, long-term solutions will increasingly involve an integrated view of resource management covering people, buildings and information technology. Building relationships across functional groups is therefore an important ingredient for a successful outcome. It is worth noting at this point, how the very term used in many businesses to define and describe organizational groupings – divisions – does itself reinforce the notion of fragmentation and separation, concepts that are outmoded in 21st century business. When developing relationship models based upon assessed interaction between functional groups, it is important to be aware that, in the vast majority of cases, the interaction most probably occurs between *some* of the personnel in each group. Further, these people, i.e. the numbers of people, may change through time so that a varying number of people from one group have varying levels of interaction with varying numbers of people in another group. Rarely, if ever, will *all* of the personnel in one group interact with *all* of the people in another group. Consequently, the co-location of people (individuals and groups) is more likely to be based upon work process criteria, i.e. some people from each relevant function being grouped as a work team, rather than all of one function being congregated together.

5.4.2 Strategies for improving space utilization

For the facilities manager, one of the principal opportunities that he or she has to demonstrate that his or her actions can add value to the business processes they are

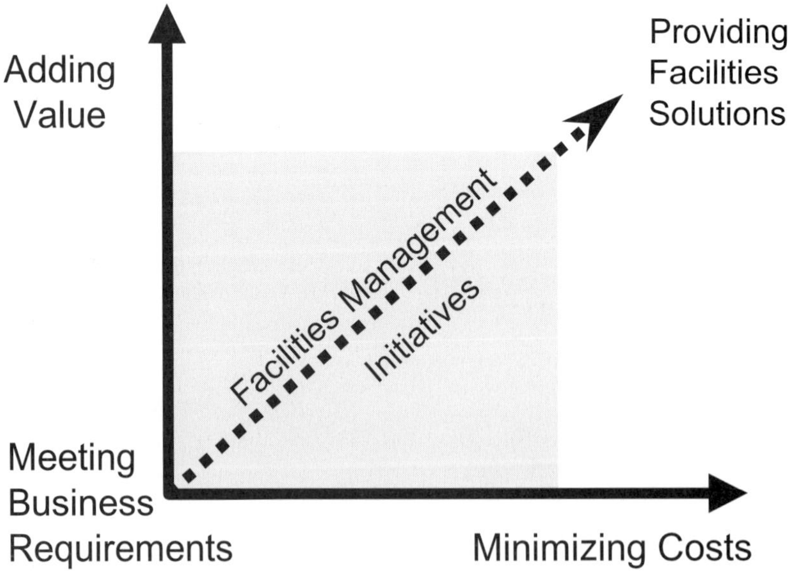

Fig. 5.13 Facilities management initiatives as a vector for change (adapted from Duffy *et al.*, 1993).

supporting, is through strategies that improve building performance by adding value as well as minimizing operating costs. This concept is aptly illustrated in Fig. 5.13 where facilities management initiatives are viewed as journeying along a vector providing facilities solutions to meet varying business requirements through time.

Facilities management initiatives aimed at improving building utilization can be considered under two broad categories:

- Those that relate to the property portfolio; and
- Those that are aimed at improving operational effectiveness.

Initiatives at the property portfolio level are mainly aimed at handling any existing, or potential, mismatch between supply of, and demand for, workspace. The motivation for change may be manifested by a corporate drive to reduce operating and occupancy costs; to realign the operational property to reflect a culture of teamwork; to relocate to less expensive, more modern, or purpose-built premises, or a combination of all of these.

Case Note 5d

Powerlink Queensland, Brisbane, Australia. Refurbishment and relocation of headquarters to a consolidated site. The drive of the new Board of Management was to change a former public sector commission responsible for the generation and transmission of power with overtly bureaucratic trappings and over 3000 staff, into a modern streamlined commercially-focused business comprising 500 employees. The site chosen for the new headquarters was a recycled 40-year-old former industrial building located on a railway yard, which was converted into a modern office building with a central atrium that incorporated water features and piped-in music, and set in a landscaped park (Fig. 5.14). Senior management saw the new workspace as a catalyst for change, in their desire to promote teamwork and the allocation of space based on function and needs, rather than status and wants (Beitz, 1997).

Fig. 5.14 An example of a standard footprint set, fronting the landscaped atrium – Powerlink Queensland.

The second group of initiatives, which are aimed at improving operational effectiveness, are mainly issues relating to changing current organizational practices. The move by some organizations to allocate workspace based on the needs of tasks rather than job grade and status can seem challenging. The introduction of desk-sharing and booking of space, are examples of fundamental mind shifts with regard to the use of workspace as a business resource rather than a 'hidden' corporate expense. The increasing shift to knowledge work, from manual, administrative and clerical work, has meant that the design of workplace configurations must accommodate the requirements of electronic information and communication tools. These prerequisites for today's businesses are factors which, for a long time, regrettably appeared to have escaped the attention of all but the most enlightened within the property industry, the evidence of which is in abundance in most cities.

Case Note 5e

PowerGen, new headquarters building, Coventry, England. The company that was formed out of the old UK Central Electricity Generating Board in 1991, developed its new headquarters building with design principles based upon the spatial integration of personnel as its guiding philosophy. The new company wished to develop its own culture and considered internal communications and team working to be important factors in achieving this objective. The aims of space planning were to minimize the effect of building constraints, promote cost effective user choice and reduce the impact of future organizational changes. The rationale governing the design of the workspace within the building was set down as ground rules (Table 5.3) that influenced the design of workplaces, location of business support centres, the use of technology and the provision of storage facilities (Allison, 1994).

It is particularly worth noting that in both the Powerlink and PowerGen projects, apart from benefiting from the strong support of senior management and their vision, the success

Table 5.3 Space-planning ground rules – PowerGen, Coventry

Ground Rules	Reason
• No furniture to be less than 600 mm from the glazed external perimeter	• Services access
• No permanently occupied workstations to be placed in reserved core areas	• Quality of environment
• Individual workstation configurations only to be selected from the footprint library	• Cost efficiency and ease of reconfigurations
• No screens or furniture in excess of 1300 mm high are to be placed parallel to the building's external perimeter	• Ventilation and views out
• No screen or furniture to exceed 1600 mm height	• Visual continuity
• Passageways between work groups will not be less than 960 mm wide	• Fire escape routes and disabled persons access
• Components must be minimized by the sharing of components in back-to-back configurations	• Cost efficiency
• Connected workstations must not be barriers to ventilation within each bay	• Ventilation

of the resulting workspaces emanates from the high functionality of the buildings, which owes much to the rigorous research and evaluation of design options, and the sensitive consultation and participation of the people in the business.

5.5 Facilities support infrastructure

The physical design of facilities, how space and equipment are allocated and used, their cost, and their relationship to organizational and technological systems are parts of an integrated whole, what Becker (1981) appropriately defines as 'organizational ecology'. Until recently, few organizations saw space as a resource that could either enhance or hinder their organizational and business objectives. Today, there is growing acknowledgement that, in order to support users to perform their tasks efficiently and effectively in a diversity of work settings, careful design, layout and management of space are essential prerequisites. In their role, facilities managers are expected to take initiatives that increase work process effectiveness, enhance individual work activities, and improve the flexibility of use of facilities. We now turn our attention to the facilities infrastructure support that is needed to promote the above initiatives in creating an appropriate organizational ecology.

All aspects of building evaluation must be considered in the context of the needs of the people in the business, and the impact of the building and its infrastructure upon them. However, some aspects of the building location, shell and skin can be assessed independently of the specific requirements of occupants, such as the generic issues of site accessibility and the adaptability of the shell and skin. The building services and technologies, on the other hand, can only be assessed in the context of how well they meet the needs of the business processes and the occupiers of the building. Table 5.4 lists the components that are typically found in commercial office buildings.

In comparison to the life cycle of the building shell and skin, the life spans of the above elements are generally much shorter, 15–20 years for major services compared with 50–60 years for the building shell. This means that building services components and technological systems will need to be replaced several times during the operational life of the building. In terms of life-cycle asset management, facilities managers must seize the opportunities for judicious upgrading or changes to the specification to meet current and projected future needs. The function of building services is to provide power, lighting, ventilation and air-conditioning or heating in order to maintain the physical environment within certain specified limits of comfort for its occupants. The choice between centralized control and local control, is an issue that has to be balanced against serviceability, and the impact of individual control on the macro environment.

Technological systems within buildings have developed in recent years, in terms of both their scope and their sophistication, made possible by the convergence of automation

Table 5.4 Building services and technological systems in buildings

Building services	Technological systems
• HVAC zoning and control	• Information technology systems
• Power provision and backup	• Communication systems
• Distribution of cabled and piped services	• Building automation systems
• Lighting systems and control	• Space management systems
• Performance monitoring systems	• Business support systems
	• Access control and security

systems and communication systems. It is an area that has spawned various studies to identify and investigate the future of what are called 'intelligent buildings'.

Building services controls have evolved from being large and cumbersome centralized systems to smaller, more flexible systems with individual controllers for separate items of plant. Areas of application include environmental services control, fire detection, security, lighting and preventative maintenance. It is through the use of such systems, for example dimmers and presence detectors, and their individual control via telephone or computer keypad, that buildings will possess the capability to be more responsive to the varied needs of occupants, an issue to which we will return in later chapters.

Case Note 5f

The British Telecom (BT) Workstyle 2000 Programme. Workstyle 2000 is a major change project that BT has been running since 1993, aimed at changing the way the people in the business work. As an integral part of reorganizing their operations bases, which involved relocating office-based staff from central London to buildings located adjacent to the M25 motorway, which circumnavigates the city, BT decided to implement a new philosophy of work which they encapsulated as 'Giving people the flexibility, knowledge and tools to determine how, where and when they work'. Many of BT's former buildings dated from the 1960s and 1970s, and did not provide an appropriate environment in terms of quality, functionality or facilities provisions for BT's needs; primarily because these buildings had been designed around out-of-date organizational assumptions, which did not facilitate good communications, nor did they have the ability to accommodate the appropriate IT infrastructure. In the new buildings, the IT infrastructure, which is an integral part of the Workstyle 2000 programme, supports both the management of the building environment and the work activities of the occupiers. The infrastructure includes networks that support the work of individuals through desktop or laptop computers at shared locations; systems that support the building's environment, heating, lighting and ventilation and control security; and systems that support the provision of facilities services in the building, such as reception, help desk, internal moves and maintenance activities (see Figs 5.15 and 5.16).

Case Note 5g

Financial services company, central England. The application of IT is already having a significant impact on the numbers of people organizations require to conduct their business. This financial services client indicated that, over a 5 year period, they are forecasting a 30% increase in the volume of customer transactions they will process, which it is envisaged will be achieved through improved IT systems requiring 50% of their current staff – i.e. achieving more with less.

Over the last decade, business support systems have grown substantially in both capacity and range for almost all types of organization. The traditional applications included in 'office automation' have given way to a whole range to tools that support business tasks today. Advanced audio-visual systems like video-conferencing, voice mail and multimedia presentations are complemented by electronic communication via Internet, Intranet and e-

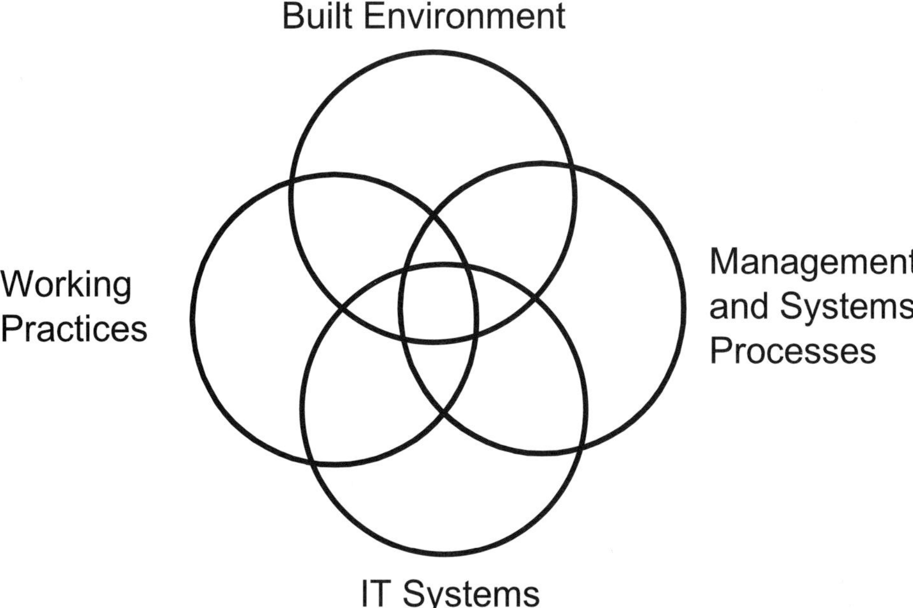

Fig. 5.15 Workstyle 2000: IT systems are an integral part of the whole (source: McLocklin *et al.*, 1995).

mail. For facilities managers, space management systems that combine the power of CAD with inventory database software have become invaluable tools for maintaining records of buildings, data about equipment and furnishings, as well as organizational data.

Other important components of facilities support infrastructure are furniture and storage systems and their adaptability for easy reconfiguration in response to the introduction of new work group layouts or work practices. They also need to be able to cope effectively with diverse IT systems while providing an ergonomic and safe environment in which the equipment can be used.

When assessing the capabilities of existing or potential workspace, facilities managers need to consider the merits of using on-site or off-site document storage for live records as well as for archives. Consideration of these issues will influence decisions about the amount of workspace it is necessary to provide for records storage, as well as the suitability of possible locations.

To summarize, in assessing the supply side of the business of space it is clear that the functionality and quality of facilities are at least as important as the amount of space to be occupied. For facilities managers, building and space audits can be extremely useful in developing a thorough understanding of the following.

- Quantitative information on the existing building stock, and other premises under consideration, which also affords the opportunity to revisit the premises policy.
- Qualitative information on the appropriateness of the buildings in meeting current and projected demands for workspace, and an opportunity to test actual space usage against an effective space budget and planning concept.
- Financial options for maintaining or modifying existing buildings, or acquiring new buildings to meet projected demand.

(a)

(b)

Fig. 5.16 (a) and (b) Responsive environments for flexible styles of working at BT's offices where it operates its Workstyle 2000.

- The need for development of an appropriate methodology and corporate guidelines, for use in making a smooth transition from a traditional supply-led provision of workspace, to an appropriately designed demand-led enabling work environment.

References

Allison, S. (1994) PowerGen's space–planning mission. *Facilities* **12** (11), 11–15.

Becker, F. (1981) *Workspace: Creating Environment in Organizations* (New York: Praeger Press).

Beitz, D. (1997) *Presentation at FMA Professional Development Program*, Brisbane, November.

DETR (Department of the Environment, Transport and the Regions) (1998) *Good Practice Guide 237: Natural Ventilation in Non-domestic Buildings* (Watford: BRESCU).

Duffy, F. (1990) Measuring building performance. *Facilities* **8** (5), 17–21.

Duffy, F., Laing, A. and Crisp, V. (1993) *The Responsible Workplace* (London: Butterworth Architecture).

Edwards, B. (1998) *Green Buildings Pay* (London: Routledge).

Flanagan, R., Norman, G., Meadows, J. and Robinson, G. (1989) *Life Cycle Costing – Theory and Practice* (London: BSP Professional Books).

Gordon, A. (1974) Architects and resource conservation. *RIBA Journal*, January, 9–11.

HOK Consulting (1993) *Alternative Officing*. The HOK Facilities Consulting Report, p. 3.

Laing, A. (1997) New patterns of work: the design of the office. In Worthington, J. (ed.) *Reinventing the Workplace* (London: Architectural Press), p. 36.

McLocklin, N., Maternaghan, M., Lowe, S. and Bevan, M. (1995) *Technology and the Workplace*. Facilities Management, Management Guide No. 9 (London: The Eclipse Group).

Then, D.S.S. (1994) People, property and technology – managing the interface. *Facilities Management*, October, p. 7.

Then, D.S.S. (1996) A study of organisational response to the management of operational property assets and facilities support services as a business resource – real estate asset management. Unpublished Thesis. Heriot-Watt University, Edinburgh, UK.

6

Matching demand and supply – the outcome

Successful space strategies resolve the needs of people in the business (the demand) with the provision of appropriate workspace (the supply). The challenge in space management is the constant adjustment of the equation between the demand for, and supply of, space. Successful space planning resolves supply and demand based upon an understanding of the present use of space and in doing so establishes a profile for the future use. The challenge is to effectively match supply to demand for space over time guided by an agreed strategic framework. Often the management reality is how to use the whole of the organization's stock of space continuously, to achieve business goals. The forecasting of a business's workspace has four elements that the facilities manager must address in attempting to reconcile supply and demand, namely:

- **Location** – the position to serve best the business's operational focus;
- **Quantity** – the amount of different types of workspace required;
- **Quality** – the functionality and standards required; and
- **Timing** – the timing and duration of the workspace requirements.

In the previous two chapters, we considered the issues relating to the assessment of space demand and the essential attributes of building supply. In this chapter we turn our attention to the issues facilities managers are likely to encounter in their efforts to maintain a constant balance between demand and supply.

6.1 Reconciling demand and supply

The basic premise that governs the reconciliation between the demand for, and supply of, workspace is appropriately paraphrased by Akhlaghi (1993) when he wrote: '. . . an organization should own, acquire, sell, maintain and adapt buildings only to the degree and in a manner which would enhance its organization/business strategy and that would serve its demand for space in qualitative and quantitative terms over pre-defined time-scales.'

From their research into the relationship between work patterns, HVAC systems and building types, Laing *et al.* (1998, pp. 15–16) described a research model in which they illustrate what they termed 'the logic of demand and supply', see Fig. 6.1. The pattern of demand of an organization is seen as progressing through several levels of impacts: from

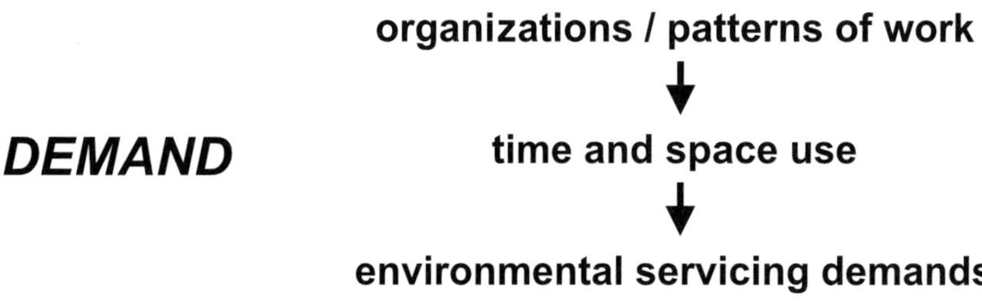

environmental systems

building constraints for space design and servicing

SUPPLY

base building types

Fig. 6.1 The logic of demand and supply (Laing *et al.*, 1998).

the pattern of work, through the pattern of time and space use, to the nature of the demand for environmental services. The demand is then contained within, and affected by, the constraints and opportunities that emanate from the hierarchy of building supply elements; the base building shell and its configuration; the limitations of space design and associated environmental servicing affinities; and the scope and performance attributes of the environmental systems themselves.

It is evident that two preconditions are required to enable these guiding principles to be put into practice.

- A strategy-driven approach to operational buildings as a supporting business resource; and
- An objective demand-led analysis of workspace.

6.1.1 Strategy-driven approach

As explained in Chapter 3, understanding the relationship between the organization and the premises it occupies is fundamental to effective space planning and management. It is important to understand space, its significance as a strategic asset as well as an operational resource and plan for its change and adaptation over time. Perhaps even more significant, is senior management's endorsement that the working environment can be used as a catalyst for change. In recent times, it is evident that trends in building use and design are more likely to be formed in response to drives for cost reductions, which almost inevitably seize on people or facilities, or both. Measures that optimize hitherto under-managed facilities resources, have been grasped by organizations in both the public and private sectors, from which the growth of facilities management as a professional discipline has been closely associated. The pressure to reduce operating costs has, in recent years, shifted from the initial focus on reducing overall staff numbers (headcount reduction) to initiatives aimed at reducing occupancy costs. The pressure to reduce premises occupancy costs has resulted in the scrutiny of two main aspects of operational property management – the amount of space occupied by business units and the utilization of existing spaces. These concerns are reflected in two research reports that focused particular attention on the future shape of the workplace and the strategic importance of effective space management for overall business success. The Gallup Report: *Shaping the Workplace for Profit* (Gallup, 1996), stresses that every business has to be as flexible as possible, needs to reduce costs and maximize productivity, and also stresses the importance for organizations in establishing a clear premises-related strategy that takes into consideration new working practices and the exploitation of new technologies. The Milliken Report: *Space Futures* (The Henley Centre, 1996) also emphasizes the need for organizations to view space management as a strategic issue that is integral to the corporate agenda.

One of the most influential publications in the late 1990s, which helped to emphasize the importance of effective management of the corporate operational assets in North America is *The Corporate Real Estate 2000, Phase One Report* (Joroff, *et al.*, 1993). It is worth noting that the Phase One report labelled real estate as the potential *fifth* corporate resource, after the other four major corporate resources of capital, people, technology and information. It is also significant that the report chooses to define corporate real estate as comprising 'land, building and work environments'. The inclusion of the work environment adds a wider dimension to the scope of real estate beyond the scope of traditional interpretation of estate management, by including issues relating to utilization of space within the corporate work environment. This view reinforces our proposition that issues relating to *facilities provision* and *facilities management* cannot be considered in isolation since they are in fact inextricably interlinked, as a corporate resource. These reports emphasize the important shift in the way operational property is being viewed by

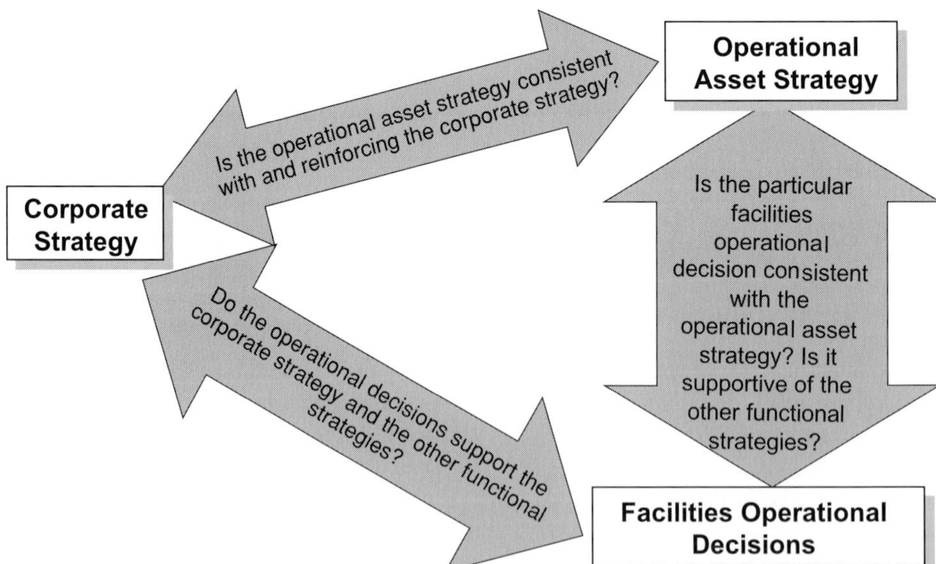

Fig. 6.2 Operational assets decisions in a strategic management context (Adapted from Nourse and Roulac, 1993).

some of the more enlightened organizations – the strategy-driven alignment of operational resources. The crucial link between corporate strategy and operational asset management in the context of strategic business management is illustrated in Fig. 6.2. It emphasizes the need for strategy alignment at two levels: (i) strategic alignment between corporate strategy and the operational asset strategy, and (ii) alignment of facilities operational decisions and the operational asset strategy.

We have previously stressed the crucial role of strategic facilities planning (SFP) in providing a mechanism for dialogue between strategic business management and operational asset management. The desired outcome is facilities managers having the capability to generate appropriate supporting facilities strategies. In promoting a strategy-driven approach to reconcile continuously the pressures of demand and supply of workspace, there must be a channel for continuous dialogue between the personnel responsible for strategic management of core business development, and their colleagues responsible for operational management of business resources. An integrated management framework that incorporates considerations of facilities provision and their ongoing management as an integral part of the business planning cycle, is clearly desirable. From research, we propose two key instruments to bring about a formal and continuous dialogue: the *Strategic Facilities Brief* (SFB) and the *Service Level Briefs* (SLB), the critical roles of which are illustrated in Fig. 6.3, in the context of the strategic facilities planning activity.

The strategic facilities brief (SFB) is the output document that defines the operational needs emanating from the organization's business plans. The principal purpose of the SFB is to define a corporate procedure that guides key facilities attributes and service performance criteria that are required to fulfil the organization's objectives as dictated by business plans. The scope of the SFB will be influenced by the following factors.

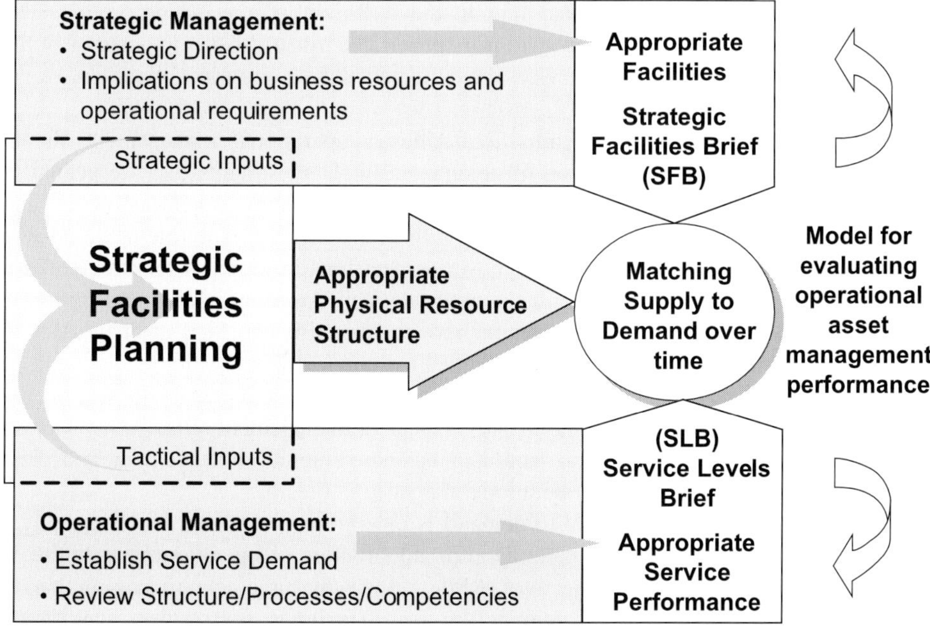

Fig. 6.3 Strategic facilities planning and the context of SFB and SLB.

- The nature of business.
- Site selection attributes.
- The need for flexibility.
- The exposure to technological change.
- The corporate view of the role of property and support services.
- Resource commitment and affordability.

The service level briefs (SLB) represent the definition of acceptable performance levels in respect of the physical asset base and the requirements for support services as defined by the SFB. The principal purpose of the SLB is to define and quantify the appropriate support services and facilities infrastructure necessary to support the activities of the business units, and their performance within the workplace environment. The scope of the SLB would be dictated by the following criteria.

- Minimizing exposure to risks within the workplace.
- Serviceability of the physical asset base.
- Protection of the asset value.
- Promotion of an environment conducive to effective working.
- An appropriate procurement strategy.
- Costs and affordability.

It is clear from the above, that the development of the SFB and the SLB involves participation and inputs from the strategic business planning process, together with contributions from operational asset management and the participation of staff. The inputs from strategic management are expressed in terms of the intentions of the corporate strategy for core business development in the short, medium and long terms. The role of the SFB is to attempt to ensure that any investments in physical resources result in the delivery of appropriate facilities that support the fulfilment of the business plans. The source of the strategic inputs is *business information*. The role of SLB is to ensure that the appropriate facilities and support services are delivered at the appropriate operational level as defined by the SFB.

Similarly, the tactical inputs from the business's operational management should contribute to the strategic evaluation of potential capital investments that are geared to altering the capacity of the physical assets, such as the building, its infrastructure and services. The source of the tactical inputs is *facilities information*. The role of the SFB is to interpret the implications of strategic business decisions in terms of requirements for operational assets, thereby ensuring that appropriate facilities are available to support the chosen strategic direction. The interactions between the SFB and SLB are necessary to ensure proper matching of demand for facilities to support the strategic development of business, with the supply of appropriate facilities and services necessary to support the implementation of the desired business plans. Proactive management of operational assets requires a clear strategic direction from the business's senior management and clear measurable deliverables from the operational management team.

The process of matching supply to demand is clearly never easy, and is frequently a complex one. The final outcome is an appropriate structure of operational assets that is aligned with current corporate objectives, while at the same time acknowledging that, at any point in time, the goal is to optimize the use of facilities and services such that a balance is achieved between demand and supply – the *current* steady state. The concept of the current steady state is central to effective management of operational assets and their associated services over time, within a dynamic business environment (see Fig. 6.7).

6.1.2 Objective analysis of space

Underpinning any development of a facilities and real estate strategy is the requirement for a robust methodology based on objective analysis. In this respect, the conventional 'supply-led' approach that focuses on building and estate management aspects must be preceded by a realistic quantification of the business's demand drivers. In other words, the matching process must be driven from *demand to supply*, not from *supply to fit*. Typically, we encounter the approach where organizations have acquired a building based on incomplete analysis and information, resulting in the users and processes having to 'fit within' the restrictive constraints of the acquired building.

An objective demand–driven process should start by determining the key requirements of the organization, which should be translated into the desired building attributes of site, shell, core, shape, infrastructure and services. Bridging the omnipresent communication gap between the guardians of business information and the custodians of building and facilities information, is fundmental to the success of the whole space planning process. It is part of a structured approach to reconciling the demand and supply factors, defined in both quantitative and qualitative terms, and ultimately reflected in the quality and effectiveness of the workplace environment and its layout.

Figure 6.4 illustrates a demand-led approach, which starts with the foundation of the *space budget* derived from objective organizational analysis of space demands. The progression from *space budget* through the intervening levels to *image and design* facilitates a logical consideration of issues that must be assessed before a supply decision is taken on any individual building, existing or new.

Figure 6.5 illustrates the process of reconciliation of demand and supply via a methodology that evaluates demand from a framework derived from business and property needs; it indicates the role of building supply specifications, current practices and the

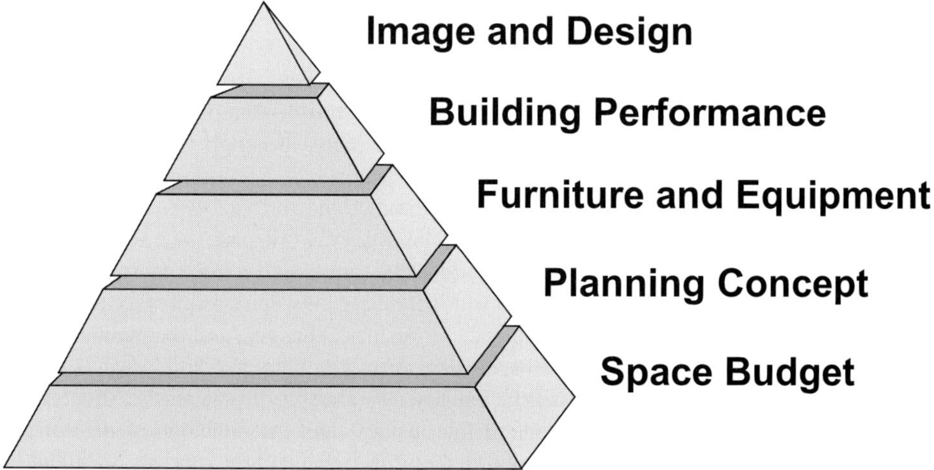

Fig. 6.4 A structured approach to matching supply to demand.

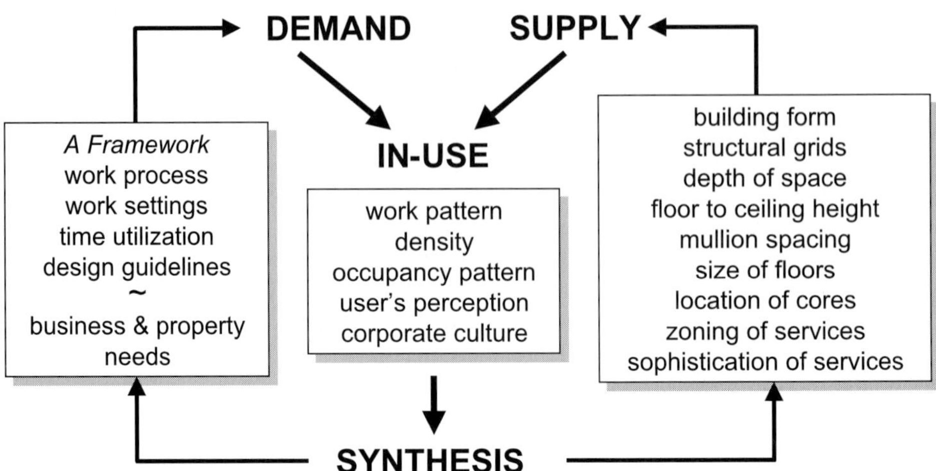

Fig. 6.5 An objective space analysis methodology.

perceptions of workspace in use (Fig. 6.6). The proposed methodology provides a structured approach to take into consideration the supply and demand factors that were reviewed in Chapters 4 and 5.

6.1.3 Handling the mismatch

Much of the guidance given to facilities managers about the provision of workspace is based upon the imperative of matching the building to the organization, although in practice it is frequently vice versa. The reality is, in our experience, that in a world where change is ever-present there will always be a mismatch. In this section we set out where and why mismatches arise, and how to deal with them. Given the volatility of the business environment, it will be an almost impossible task for any facilities manager to claim that they have managed a situation in which there is no mismatch between demand and supply. Where this situation is achieved, experience suggests that it will be short-lived. In practice, it is a matter of assessing the degree of mismatch that exists and determining if it is acceptable relative to the type of business activities that are to be supported.

In space planning and management, mismatch between demand and supply is essentially a by-product of the timing of need and availability. As a product, the delivery of fully serviced space has rarely if ever been immediate, although the availability of commercial fully-serviced offices, call centres and factories is becoming more common. For most organizations it is reasonable to expect some lead-time between requests from business units for additional space and the availability of workspace in a fully serviced state. For a newly constructed building, the delivery time could extend to several years from planning to occupancy. In the main, it is this time-lag between a request for additional or different functional space as a consequence of business changes, and the availability of appropriate workspace that is responsible for the mismatch between demand and supply. Rarely does the mismatch occur instantly, usually it is a result of progressive divergence between the changing operational needs of the business and the unchanging state of the premises and its infrastructure. Consequently, facilities managers must be ever-vigilant in monitoring the situation, so as to detect the early signs of progressive performance decline

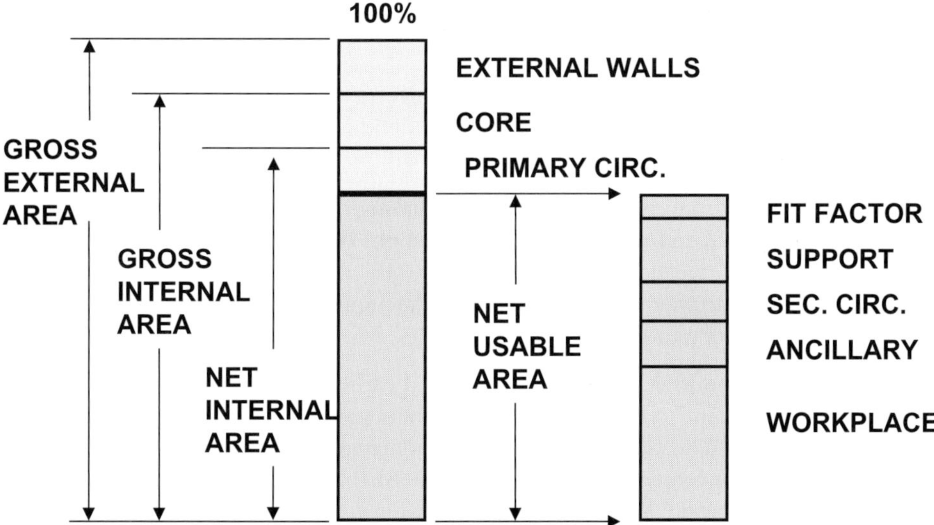

Fig. 6.6 The elements of workspace area.

before it reaches sub-optimal levels with a corresponding impact upon business performance. All too often the reason for mismatch and the speed at which it arrives at the facilities manager's door, is a direct consequence of the absence of an appropriate dialogue with business managers. In turn this frequently leads to what we call 'the Friday afternoon syndrome' in which the facilities manager receives a telephone call from the business manager advising that, 'I have five people starting on Monday, where will they be seated?'. So, for all sorts of reasons, there will almost always be a degree of mismatch at any point in time given the pace of change all organizations are confronting. As a business resource, mismatch between demand and supply of workspace can be viewed from a number of perspectives and at various levels. Fundamentally, the nature of a mismatch can be categorized as either strategic or tactical. A strategic assessment will involve a macro-level analysis of the corporate building portfolio aimed at identifying mismatch between demand and supply as part of the strategic facilities planning process, involving the scenario modelling referred to in Chapter 3. Such an assessment is likely to include the following.

- Mismatch arising from an unacceptable amount of space that is surplus to business requirements, across the whole operational asset portfolio.

- Mismatch arising from uneven distribution of the appropriate types and mix of workspace across the whole portfolio, often resulting in 'pockets' of surplus space and/or unsatisfied demand scattered across the portfolio.

- Mismatch arising from having a poor balance between types of building and/or facilities and the requirements of the business, either in terms of inappropriate physical conditions and/or environmental servicing of workspaces (technical obsolescence); inappropriate location in relation to resources availability, e.g. workforce, raw materials or markets (functional obsolescence); or an unacceptably high cost of upgrading to required operational levels (economic obsolescence).

- Mismatch between the current tenure and other forms of provision, e.g. direct ownership, leasing, or sale and leaseback.

At the next level, a tactical assessment will involve a micro-level analysis aimed at identifying mismatch that arises at the level of individual buildings and within business divisions or functional units. Such an assessment will include the following.

- Mismatch arising from an imbalance between timing of need and timing of delivery.
- Mismatch of spatial and service quality expectations of business units and their supply by the facilities providers.
- Mismatch arising from uneven distribution of the required types and mix of workspace within a building or part of a building.

The concepts of the SFB and the SLB described earlier (Fig. 6.3) should be seen very much as the 'lubricant' within an integrated framework for promoting an informed dialogue between business managers and facilities managers; a concept we will return to in Chapter 8. The convergence of the SFB and the SLB offers a way for reconciling any imbalance between supply and demand at three levels.

- First, reconciliation of demand and supply within the domain of the SFB, when operational demand drivers are resolved by identifying the most appropriate real estate variables to modify.
- Second, reconciliation of demand and supply within the domain of SLB when facilities and environmental support demand drivers are resolved by identifying the most appropriate facilities services variables to modify.
- Third, where the required enabling work environment is derived from the resolution of a combination of variables identified from both the SFB and the SLB.

Figure 6.7 illustrates how facilities managers can play a key role in resolving the constant tensions between demand and supply by skilfully managing the information flow

Fig. 6.7 Reconciling supply and demand to create an enabling workplace environment (Then, 1996).

from the business domain, and applying building and operational knowledge to providing continuously appropriate facilities solutions to meet business requirements.

The emphasis on 'adjustments to the next steady state' recognizes the volatile environment of the business world in which facilities managers must operate as proactive agents for change. The context of the different management focuses of 'managing alignment' and 'ongoing management' within the respective domains of the strategic facilities brief and service levels brief is important, as both the SFB and the SLB are driven by different sets of factors, which are set within the constraints of the corporate culture in which they must operate.

Whilst facilities managers cannot realistically hope for an optimum balance at all times, it makes good practical sense to prepare for the worst case scenario by taking measures to mitigate against a potential level of mismatch between demand for, and supply of, workspace that is economically unsustainable for the business. Measures that can be taken to mitigate the magnitude of mismatch, must feature at both strategic and tactical levels.

At the strategic level:

- Strategic business planning must include procedures that incorporate the facilities implications of business initiatives – linking demand for space to projections of business plans; that is, positive or negative growth. From our experience, introducing change at core business level is both an issue of perception and awareness of the contributory role of operational assets to the financial success of the organization.
- Facilities managers, for their part, need to understand the nature of the business and its demand profile and have a thorough knowledge of the constraints, as well as the opportunities (flexibility – short-term and adaptability – long term) of the existing operational portfolio.

At the tactical level:

- Facilities managers, in collaboration with business users of workspace must consciously plan ahead to establish the appropriate level of flexibility in supply, as well as appropriate disposal or exit strategies; for example, facilities provision and ownership issues, lease arrangements, and potential arrangements for sharing workspace.
- Facilities managers need to establish clear guidelines for how the enabling workplace environment addresses progressive changes in corporate culture, space and workplace provision, and amenity levels.

Case Note 6a

Manufacturing company, Fife, Scotland. Ten years after opening premises that were both their factory, administration and sales office, this manufacturer had redeveloped the workspace wholly as offices. When the premises were first occupied, approximately 95% of the workspace was factory, accommodating machine shops, assembly lines and materials stores, with the remaining 5% allocated to enclosed offices. As the business expanded, creating demands for increased production, new state-of-the-art machinery was acquired which brought with it demands for more space and more sophisticated services, both of which could only be addressed by new premises. As a new factory was brought into operation the original premises were released and progressively adapted for use as office space to accommodate the burgeoning telesales operations, which were driving business growth to an annual turnover of £40 million, as well as larger administrative support teams.

The adaptation of the factory was carried out in a piecemeal manner as parcels of workspace were released, resulting in a disjointed group of spaces that bore little, if any, relation to the needs of the people who were accommodated within them or their work processes. And so it would most probably have remained had it not been for a major fire that completely gutted the premises. But rather than immediately embark upon rebuilding what had previously existed, the chief executive realized that the fire, although highly disruptive for the business, presented an opportunity to address many of the failings previously identified in the workspace but which had, until then, not been considered practical to address. These included: reorientation of key functions in the building by repositioning the entrance and access to the building, creating a large open work area at the centre of the building and a range of large and small enclosed spaces, as well as increasing the overall amount of workspace by extending the building, and improving the services infrastructure. New premises were available for use 23 weeks after the fire, with staff working in a functional and supportive environment, which only a short time before had seemed a distant prospect. A clear case of turning adversity to advantage.

Planning for expansion

Having determined that the business has a requirement for additional workspace to cope with expansion, the facilities manager then has to consider how it is best accommodated. Generally, where the buildings currently occupied possess the required capacity to meet the future needs, the options the facilities manager faces are either:

- Option One – should the workspace be planned to meet the current business requirements and then re-planned when the need for expansion arises?
- Option Two – should the amount of workspace assessed for future needs, be zoned as part of the current workplace layout planning, but not fitted-out with furniture and equipment until the appropriate time?

In the first option the benefits of tighter control of the workspace have to be weighed against the likelihood of greater disruption to occupiers when the expansion space is brought into active use. It will not just be the groups requiring the additional space that will be disrupted, but all of those occupying the adjacent workspace will also suffer accordingly. With the second option, the *ripple effect* of re-planning adjoining workspace can be minimized if not avoided. However, in selecting this approach, the facilities manager needs to assess the impact that the vacant space may have upon the overall ambience of the work environment, in terms of its visual appearance. Empty, and especially unlit, workspace can, unfortunately, be a depressing influence upon the occupants of the surrounding areas. In addition, consideration should also be given to the practicality of preventing 'unauthorized' use of the vacant workspace, which can take many forms including 'encampment' by both temporary and permanent workers, and a 'dumping ground' for unwanted or surplus products or materials. Although such types of ad hoc use are not consistent with good practice in the management of workspace, the facilities manager should however note them, as they are likely to be indicators of other issues requiring attention.

Of course there will be many businesses where the perceived luxury of providing today for the needs of tomorrow, will not be an option afforded to the facilities manager, either because the amount of workspace is not available or because the business believes that the cost of unoccupied workspace cannot be justified. Such a corporate mindset can give rise to what some might call, option three, that is to address the provision of additional workspace, 'when the situation arises, and not before'. As readers hopefully will be aware by this point in the book, such a reactive approach to the management of workspace is not one we would countenance. It carries high risks and is potentially the most disruptive to the business, resulting in it frequently being the most costly option of all. However where provision of workspace is made against forecast expansion, the facilities manager needs to consider the following issues.

- Should the space be allocated now to the work group that forecast the need for future expansion space?
- Should the expansion space be grouped together such that it provides a pool of space from which the facilities manager can draw as business needs dictate, whether or not it was forecast?
- If the space is allocated to the work group that forecast it, but they then subsequently no longer require the space, what will be the impact of re-planning the affected area to liberate the redundant space for the use of other occupiers?
- How should the expansion space be configured such that it is not used in the intervening period unless a valid business reason arises?

The answers to these questions are site specific, and hence can only be addressed in the context of each business's workspace and the way it is occupied. In pondering the approach that is the most appropriate in the circumstances that prevail, we suggest the facilities manager is mindful of the old saying 'you never miss what you never had', which is certainly true where workspace is concerned. The corollary is however also true, in that it is hard to give up something you have become used to having, which again experience shows is especially true with over-generous workspace provision. When assessing the business's requirements for expansion space, the facilities manager must give thought to the following.

- Are the forecasts by business managers, upon which the workspace projections will be based, well founded? How accurate have their previous forecasts proved to be, and if they have been significantly adrift of the actual requirement (either over or under) to what is the difference attributed?
- When is the requirement forecast and for how long is the workspace needed? Is the forecast for a single increment in workspace, or is it a series of additions? What the facilities manager should aim to identify is whether the forecast is a temporary blip or a long-term requirement, as both needs could be addressed quite differently. In this context it is important to be clear about what is meant by long-term, since for a business manager it is likely to be a much shorter period (3 to 5 years) than the facilities manager would be accustomed to when dealing with property leases of 10, 15 and 25 years duration.

- Where the forecast in demand is based upon a gradual increase in the business's needs, then what is the capacity of the existing workspace to accommodate the additional requirements until a sufficient volume of demand necessitates implementing a different approach? For example, the addition of one person, or a single PC would not of itself justify securing additional workspace. However, the addition of 100 personnel or a large item of equipment is likely only to be accommodated by the provision of additional workspace.

In pondering this aspect of the reconciliation of demand and supply, it is easy to see how the facilities manager's mind could turn to the proverb about straw on a camel's back! However, it is through the application of scenario modelling techniques, referred to in Chapter 3, that facilities managers will find the key to acceptable solutions to these issues.

Ways of providing workspace

Investment in work buildings and their infrastructure should be considered in much the same way as investment in manufacturing plant, machinery and equipment. Workspace should be assessed in terms of its contribution to the productivity of the business and the level of return that will be provided on any investment in it.

A key aspect of the role of the facilities manager, is to ensure adequate and appropriate investment is made in the business's workspace and that the building and its infrastructure are maintained and operated at optimum levels of performance commensurate with the needs of the business. However, as previously indicated, business needs are not a static set of requirements, but are continuously evolving; therefore, as the nature of work changes so too do the attributes required of each work setting. In assessing operational space demand at the portfolio level, Bon (1989) advocates a space accounting and planning system that captures changes over time. 'Useful space' is seen as a scarce resource available to an organization. In this respect, space needs (that is, demand for building-related services) must be determined on the basis of a plan. At a given point in time, any unit area of space used for the organization's operations can be said to be in one of nine

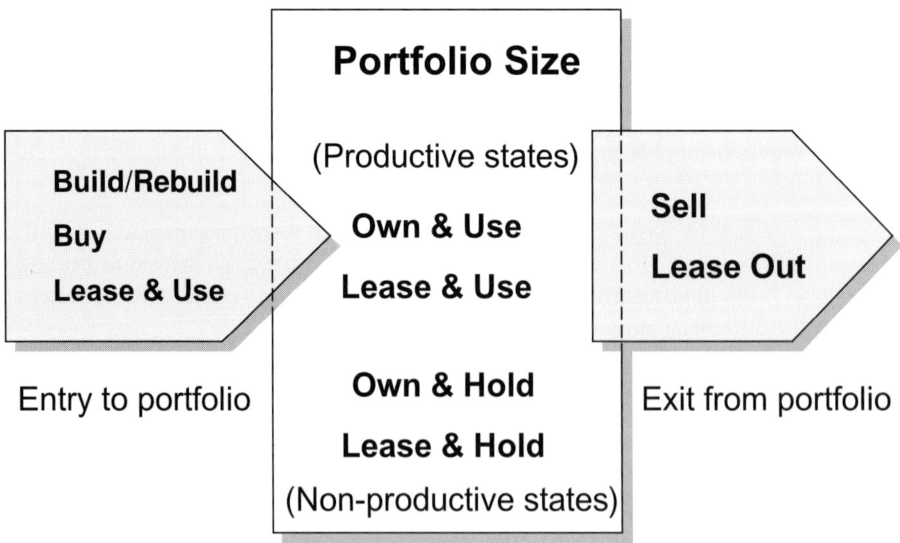

Fig. 6.8 Determinants of real estate portfolio size.

states: build, rebuild, own and hold, own and use, lease out, sell, buy, lease and use, and lease and hold, as illustrated in Fig. 6.8 (adapted from Bon, 1989, pp. 120–121).

Workspace may enter the portfolio via three states, 'build', 'buy' and 'lease and use' and it may leave the portfolio via two states, 'sell' and 'lease out'. These five states attract most attention from top-level executives of the business, as they affect the portfolio size. When considering the organization's use of space, only two states, 'own and use' and 'lease and use' are productive. The space accounting system described by Bon (1989, pp. 120–123), with its emphasis on the choice of transition paths, provides a useful tool for monitoring changes in demand over time. The stream of resources available to an organization will determine the maximum rate at which transitions from one state to another can be made, that is the supply of building-related services over time. Transitions of state can also be represented along the time dimension. This framework provides a sound basis for evaluating the consequences of alternative scenarios for an organization's operational property portfolio and the required managerial actions aimed at matching supply to demand by aligning the operational asset strategy to the organization's business strategy.

At the portfolio level, one of the key concerns of managing the corporate operational asset base is the maintenance of a continual capability to adapt to changing economic conditions; that is, to ensure alignment and relevance with the strategic direction of the business. The management of adaptability and flexibility of the corporate building stock is essentially one of managing the speed of response in supply and/or disposal of the built assets. In this respect, it is essential to take a 'resource' view of the entire corporate building stock when evaluating strategic options or choices. Bon (1989, p. 122–125) suggests that issues relating to the management of adaptability and flexibility can be viewed at three different management levels: that is, the management of individual buildings, a cluster or group of buildings, and the portfolio as a whole, as well as from the three distinctly different aspects of management – physical, organizational and financial. Table 6.1 summarizes the focus of managerial attention at each level.

Table 6.1 Management of adaptability and flexibility (Bon, 1989b, pp. 122–125)

	Aspects		
Management level	Physical Issues	Organizational Issues	Financial Issues
Building level	• Physical forms and building attributes, e.g. column spacing, number of floors, etc.		
Cluster/Group level		• Overall spatial distribution of departments/divisions within buildings and site • Site layout, physical adjacencies, expansion potential	
Portfolio level			• Appropriate mix of owned and leased facilities in response to changes in economic conditions • Geographical distribution of facilities

Shaded areas indicate predominant issues

6.2 Developing types of organizational layouts

The layout and structuring of workspace is based upon a balance between corporate objectives and individual aspirations. A successful layout will have to satisfy:

- The pattern of work to be undertaken;
- The importance of individual privacy and/or group interaction;
- The probability and speed of change;
- The control of adaptation costs;
- The take-up of IT and other technologies.

6.2.1 Categorization of workplaces

Workplaces can be categorized in four distinctly different ways. First, in relation to the way they are formed, most significantly in terms of the degree of enclosure they are afforded. Second, by the way they are used, in terms of the type of activities they have been designed to support, such as solitary or collaborative working. Third, by 'ownership', that is the degree to which they are dedicated to one or more people, or shared by many. And fourth, by the pattern of their use, be it continuous or intermittent.

Enclosure-based workplaces

- **Fully enclosed** – a room formed by floor to ceiling partitioning with access by means of a door, can provide a high degree of privacy, security and confidentiality.

- **Open-plan** – each workplace may be separated from its neighbour by means of low-level screening, or with no means of enclosure whatever; ideal for large information processing or project-based groups who need to work closely together on common tasks requiring the sharing of information; affords easy contact, but no privacy; may be landscaped or clustered to provide group identity.

- **Combi** – as its name suggests, this type of workplace is a combination of enclosed and open-plan, where the enclosure is full height on three sides with the fourth side completely open, which is ideal for distraction-free working, that does not have a need for privacy and/or security.

Ownership-based workplaces

- **Dedicated** – the workplace is allocated on a 'permanent' basis to a single individual for his or her exclusive use.

- **Shared: bookable** – the workplace is shared by many people, each of whom is able to book the workplace for periods during which it is available to them for their exclusive use.

- **Shared: available on demand** – the workplace is shared by many people on a 'first come, first served' basis, at which time it is then available to the occupier for exclusive use.

Activity-based workplaces

- **Solitary** – the activities are carried out by an individual on their own.
- **Collaborative** – the activities require the interaction of two or more people, each of whom may participate in the work process at different levels.
- **Quiet** – the activities are dependent upon distraction-free working.
- **Communicative** – the activities involve the use of communication systems and techniques, audible and/or visual.

Time-based workplaces

- **Continuous** – the operations are, for all practical purposes, ongoing throughout the working day.
- **Intermittent** – activities are carried out at various times and for various durations.
- **Touchdown** – activities are almost exclusively of very short duration.

In practice of course, every workplace is fashioned out of a combination of these four elements.

Providing common support facilities can be equally challenging. Although there is a demand for these central facilities, is the demand sufficient to warrant provision of the facilities? In many work buildings, the space with the poorest level of utilization can be the restaurant, catering for a lunch service that spans 2–2½ hours in the middle of the working day, and then lies empty for the remainder of the time. Through improvements in layout, the provision of appropriate support services (including freshly brewed coffee) and by active promotion, increasing numbers of businesses are reaping benefits from using their restaurant facilities as informal meeting spaces (Fig. 6.9).

Consideration of common support spaces will also include issues about the amount of workspace it is necessary to provide for records storage, as well as the suitability of possible locations. Cost benefit studies should be performed to evaluate the feasibility of using electronic storage systems with their much reduced requirement for workspace in favour of retaining 'space hungry' hardcopy records. The cost effectiveness of converting existing records into electronic form needs to be evaluated in relation to the potential saving of workspace and associated servicing costs that would result, taking account also of the likely frequency of access that may be required to the records. It may be feasible to convert records to electronic storage as they are accessed, i.e. as they are returned following retrieval from hardcopy archives. However, this is likely to result in a prolonged period during which records would be split between hardcopy and electronic media, which may be operationally disadvantageous. Document image processing (DIP), also called work flow management systems, has the potential to make important contributions to work process improvements, particularly through speeding up the process of accessing records. These systems can also make major contributions to reducing the business's need to store paper records. Some businesses (chemical processing and oil industries amongst others), because of the nature of their work, have statutory or legal obligations that require them to retain indefinitely certain documents (drawings, reports, specifications and the like) in their original hardcopy form, i.e. not

Fig. 6.9 Using the company restaurant for informal meetings – Marks & Spencer Financial Services, Chester.

photo-reduced, microfilmed or electronically scanned. Consequently, the workspace required to accommodate such needs has to take account of the likely future growth in such records, in terms of quantity and timing.

Computer training requires the right equipment in an easily accessible environment, which is not always available in large companies, and rarely found in small ones. Where such dedicated facilities are provided, their location is not always suitable for all staff, and the cost benefit can, and often is, questioned when they are standing idle. It is just such an eventuality that Spring Mobile Training, in Hampshire, UK seeks to address. Using purpose-built and equipped coaches, Spring provides mobile facilities for training in PC and software applications, for organizations that include the British Army, NatWest Markets, Siemens Nixdorf and Tesco (Fig. 6.10). For businesses such as these, with multi-site estates, comprising buildings of various sizes and types, hiring mobile facilities could be an extremely cost-effective alternative to fixed common support services.

The workspaces of businesses that adopt remote forms of working practices, are likely to incorporate a higher proportion of common support spaces and facilities than more traditionally based organizations. When remote workers come into the 'hub' or 'central' workspace it is most likely to be to interact with others or to use facilities that are not available to them elsewhere. Consequently, instead of the preponderance of workspace (typically 70%) being allocated to desk-based workplaces, the majority (over 60%) could comprise common support spaces such as meeting and presentation rooms, project spaces and collaborative work settings.

Most often when workspace is discussed, the talk is of the quantity, but seldom of the quality. Functional quality may be, but rarely if ever is, the aesthetic quality discussed. 'Boring places breed boring thoughts and boring people', so says Charles Handy (1997).

(a)

(b)

Fig. 6.10 (a) and (b) Mobile training facilities, as supplied by Spring, can avoid the need to provide dedicated facilities that may not be fully utilized.

He continues, 'Quality and style [in the workplace] however, encourage quality and style in their inhabitants'. The linkages between the visual quality of the workplace, and the morale and hence performance of its users, is a vital factor that the facilities manager must recognize in the design, planning and maintenance of workspace.

6.2.2 Using space standards

Workplace standards can be developed from the space required to undertake particular job functions satisfactorily. The workplace, whether open or enclosed, comprises the elements of work surface, equipment, storage, seating area and intra-circulation. The area required for each task or activity is built up from an analysis of the space required for individual elements of the task, derived from examination of the processes involved, research and experience. To be effective, workplace standards must be founded upon objective requirements and not impressions of needs and subjective desires. As jobs increasingly involve many tasks, several of which may have potentially conflicting requirements (such as quiet and collaborative activities), then the capability of any one workplace to satisfy the needs of all tasks is highly improbable. Consequently, space standards should be considered as guidelines for *taskplaces* rather than definitive allocations of workspace to an individual. This is a subject to which we will return later in this chapter.

Where space standards are developed, we recommend that each space standard is illustrated in plan and three-dimensional view and is supported by a schedule of the components it contains, comprising furniture, equipment and circulation. The use of such a tool, in a hardcopy manual or in electronic form, can be of benefit in administering the allocation of space on a consistent basis, as well as an aid to layout planning and workspace management (see Fig. 6.11).

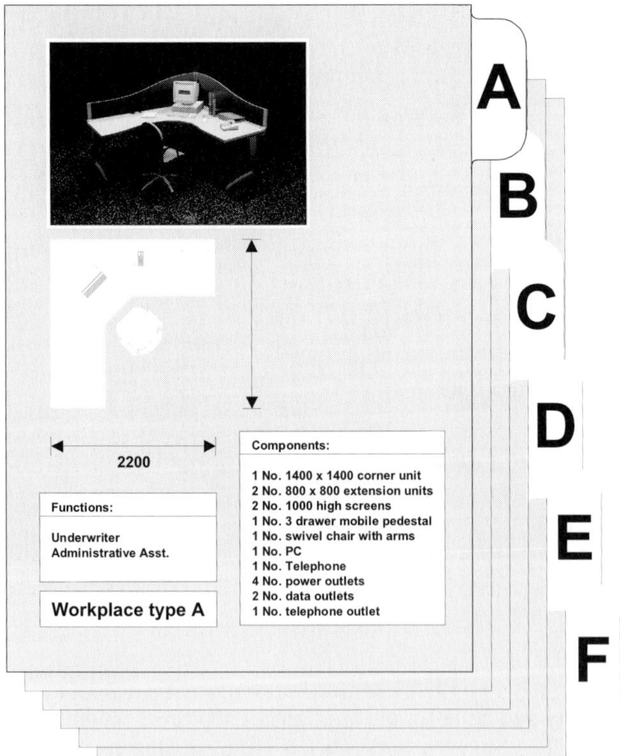

Fig. 6.11 Space standards can be of use for allocating workspace, as well as for detailed workplace layout planning and ongoing management of workspace.

We have seen several examples of businesses adopting a standard workplace provision, where a universal workplace configuration is provided in a fixed layout intended to meet the needs of all workers. Unless each worker is performing the same tasks it is unlikely that such an arrangement will find much success. This conclusion should not be too surprising, since experience has shown that rather than being founded upon a desire to meet the needs of workers, often the intention of such an approach is to make life easier for the facilities manager, by reducing the amount of workplace moves he or she has to undertake. However, we have never yet encountered a situation where ignoring the true needs of users made life easier for anyone! As Becker expresses it 'standardise on the principles and processes used, not the footprints' (Becker and Steele, 1995, p.149).

6.3 Provisions for flexibility

It has been said increasingly frequently as we have progressed through the 1990s, that the demand for headquarters buildings is on the decline. Although we do not subscribe to this view, that is not to say that we are expecting headquarters buildings to remain unchanged. On the contrary, as headquarters buildings are a manifestation of the business, then as the organization of business changes so too will its workspace. While we readily accept that there will continue to be businesses that wish or need to develop edifices as expressions of their corporate egos, these we believe will be few in number. Instead, most businesses will focus upon creating environments that are effective places in which to work, and that possess capabilities, the application of which will enable the business to minimize its exposure to fixed costs. In such a regime the true ethos of the business will shine through.

A good example of just such an approach is the European headquarters of Microsoft in Thames Valley Park, Reading, UK. When planning their new facility, which they occupied in 1997, Microsoft decided to develop the building as a series of pavilions within a campus, rather than as one monolithic building (see Fig. 6.12). The campus is planned as a series of six linked pavilions, three of which constituted the first phase with the remaining three pavilions affording some future flexibility in terms of space for expansion. However, it was not only expansion Microsoft was seeking to address by this approach. Each of the pavilions is designed as a self-contained building that could be 'disconnected' from its neighbours and released for occupation by others, in the event that Microsoft's needs at Reading reduce. This is not to suggest that Microsoft is anything other than confident about the future of its business. On the contrary, by adopting such an approach they can remain confident that they have sufficient flexibility to cope with changes in their marketplace (Fig. 6.13).

A similar approach was adopted by the Nationwide Building Society, when it developed its head office in Swindon in 1991, as a group of four blocks each of three storeys linked by an atrium that acts as a covered street between the blocks; which are also connected by high-level bridges (see Fig. 6.14). Both of these illustrations are examples of organizations that recognize that the requirement for workspace at the centre of their business can decrease as well as increase, as they adapt to different ways of meeting the needs of their customers.

Provisions for flexibility can also be made through incorporating into the premises features that afford variety in the way they could be used. The studies of the Learning

(a)

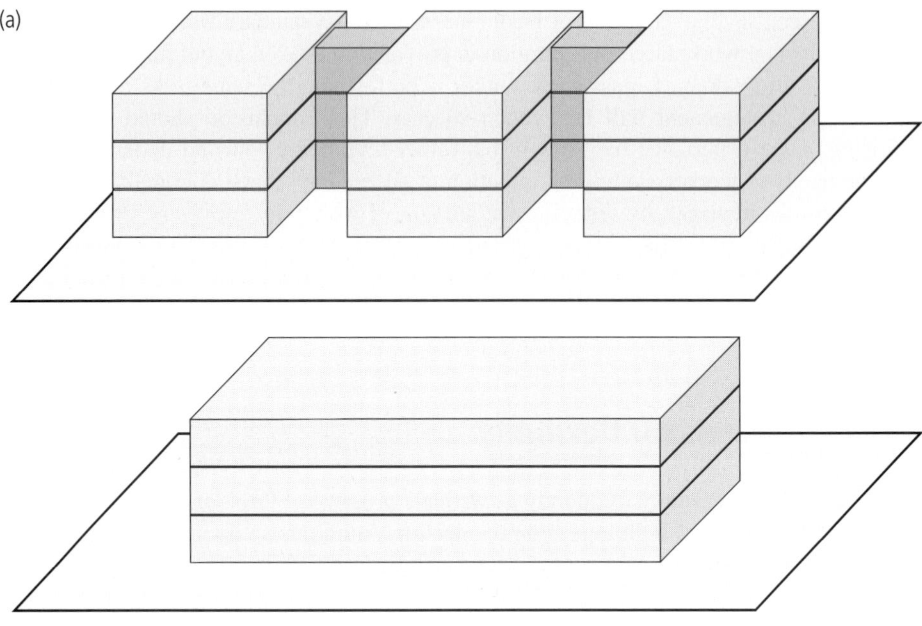

(b)

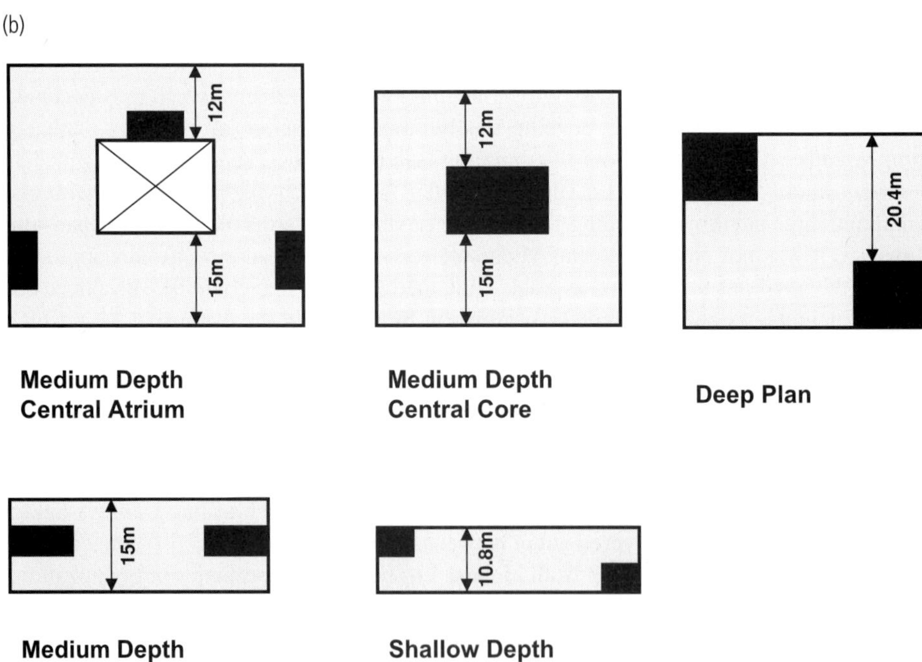

Fig. 6.12 (a) Developing workspace as a series of self-contained but linked pavilions affords significant flexibility over a single monolithic building of the same size. (b) Different core locations and workspace depths.

Fig. 6.13 Microsoft's European Headquarters, Reading, demonstrating a modular pavilion approach.

Fig. 6.14 Nationwide House, Swindon, the head office of the Nationwide Building Society, demonstrates a modular approach.

Building Group (LBG) led to proposals being developed for best practice in the design and development of workspace (McGregor, 1994). The LBG defined a learning building as one that adds value and is timeless, through taking a holistic view of its design, production, and procurement processes as well as the needs of its lifetime occupants. Learning Buildings possess the following five key characteristics.

- **Adaptability** – Ensuring the environment, both internal and external, can be configured and re-configured to suit different building users, their changing needs, work processes and layouts.
- **Capability** – Providing the potential to introduce, replace and change building elements, services and systems throughout any user's occupancy of the building and the building's life.
- **Compatibility** – Ensuring that all aspects of the building are wholly coordinated and integrated, and none are selected or installed without their impact upon, and the influences from, all other elements being considered.
- **Controllability** – Providing users with the means to maximize their use and operation of the building, its services and facilities, while minimizing the conflicts between corporate values and individual values.
- **Sustainability** – To ensure that the building and its facilities as 'assets' are operated and maintained to enhance individual and corporate productivity, their health and well being at all times, and environmental responsibility throughout the entire life of the building.

With the changing pattern of work resulting from increases in automation and Information Technology systems, the office should be, as a Robert Propst (1968) expressed it in the far-sighted publication of the same name, 'a facility based on change'. To address this objective successfully, workspace – office and non-office – must be responsive to change over time as well as making provisions for the differing life cycles expected of its elements, such as the building envelope, plant and machinery, information services, furniture, equipment and decoration. The degree to which workspace achieves these goals and is adaptable through time, is dependent upon the extent to which provisions for future flexibility were considered and accommodated in the initial plans for the premises, as referred to in Chapter 3.

Where work was previously paper-based it is likely that it may now be increasingly electronic-based, or at the very least a combination of both. As businesses and their work processes evolve from one state towards the other, then workplace settings also need to change. For example, in a call centre, as paper-based reference material is replaced by screen-based references, then the workplace footprint may be reduced in line with the reduction in the need for reference space. Such migrations in information storage media from hardcopy to electronic will also have a major impact upon the amount and type of storage facilities needed by the business, with a corresponding reduction in the amount of real estate necessary to accommodate it.

Flexibility has been, and no doubt will be in the future, a much misused, abused and poorly understood term. It falls to the facilities manager to define the constituents of *flexibility* in the specific context of his or her business operations. The wide interpretation that can be applied to the term has led to many instances where facilities have been acquired and installed in the name of flexibility but have never been used, or their potential to add value to the business never realized – for example, the installation of relocatable partitioning, the cost justification of which was most probably based upon the ease with which the enclosures so formed could be reconfigured, but which has remained unmoved for many years, possibly for as long as the natural life of the products themselves. The initial investment in these partitions only becomes a cost-benefit through time, and increasingly so, the more frequently the enclosures are moved. Therefore, where they are not moved, or moved infrequently, their full benefit is never realized. That is not to suggest that enclosures should be moved just to support some previously made justification, as part of some post-event rationalization. But rather, by giving due consideration to the likely drivers for change in the context of an in-depth understanding of the capabilities of the business's premises, the facilities manager and his or her workspace designers can specify the most appropriate components accordingly. We have, however, heard some designers retrospectively refer to provisions for flexibility that were never used, as being *insurance* against the need for change. While such an assertion may be valid, the facilities manager must always aim to balance the cost of insurance with the scale of the assessed 'risk'.

High density 'hot spots' requiring enhanced ventilation and possibly cooling, can result locally where very high concentrations of people and workplaces are created. For example, in call centres, the use of smaller than average footprints, each with its own heat generating equipment, can create a very oppressive environment (see Fig. 6.15). Therefore, in addition to creating flexibility for workplace layout planning, the design of such buildings should incorporate flexible environmental services that will enable potential 'hot spots' to be suitably tempered.

The pace of change of work activities has led to a phenomenon not previously encountered, namely work processes that have a shorter life than furniture and other

Fig. 6.15 The high density of technology within call centres can pose challenges for environmental systems – Nationwide Building Society, Northampton.

workplace products employed in their execution. As yet, few manufacturers in the furniture industry appear to have recognized this occurrence, let alone developed products to address it. The provision of workplaces should therefore be seen as a corollary of business itself – where the diversity of workplaces should be seen as a market differentiator in segregating the effective organizations from the ineffective. What is essential, is that we create what Becker and Steele (1995, p.194) describe as 'people places not object spaces'. However, when developing people places care must be taken to achieve the correct balance. 'Them and us' environments, although an expression of what are becoming increasingly outmoded styles of management, are regrettably still practised by some businesses. Such environments are characterized by more time, effort and money being spent on the workplaces of senior management than on other members of the organization. An equally jaundiced expression of this approach is evident in the marked difference taken to the visual appearance and quality of environment that visitors to the premises will encounter in preference to those areas used solely by staff. Putting these examples together, one might almost say that the way some organizations handle the design of their workspace (in terms of effort and resources expended on it) is in inverse proportion to the amount of time the 'recipient' will spend in the building, i.e. visitor areas get most attention, followed by senior management areas and lastly the spaces occupied by the resident staff population. Too frequently, designers see the end-product of their labours as being a monument to their 'design ability' rather than a work environment that truly supports the work needs of people. The irony is, that if the environment was truly supportive of its users, then of course it would stand as a testament to the capabilities of the designer – no finer testament could surely be appropriate.

6.3.1 Ensuring business continuity

As referred to in Chapter 4, it is of vital importance in developing an accommodation plan that the organization takes due account of its needs for business continuity and disaster recovery. This is likely to involve both active and passive measures, including assessing the following.

- What impact will these requirements have for the amount of workspace the business should provide, its location and its infrastructure?
- What percentage provision should be made for business critical activities as well as for other functions and work processes?
- For organizations that have more than one site – to what extent can each site, individually and/or collectively, support the needs of the business in the event of the loss of another business location?
- Are there facilities available within an acceptable distance of the business's workspace that could be hired when required? And if not, could contingency workspace and facilities be shared with other businesses – partners, alliances, suppliers and even customers?

The facilities manager needs to explore these issues individually with each work group and then consider the combined effects for the organization as a whole. The strength of a business continuity plan lies in taking a holistic approach to the composite requirements of the entire organization, rather than the recovery ideals of just one department or work group, no matter how crucial.

6.3.2 Keeping up with technological development

As technology changes then workspace and the way it is designed and equipped needs to change. The technology of the workplace continues to develop at phenomenal speeds. The most significant amongst the many areas that will impact upon the design and use of workspace, include the following.

- Convergence of technology means more and more functions are being incorporated into a single desktop PC; this, together with multi-function devices that incorporate printer, fax, scanner and copier, will result in fewer discrete devices being required in the workplace.
- Developments in battery technology mean that laptop, palmtop and other similar equipment do not need constant connection to mains power.
- Cordless LANs and telephones reduce the intrusiveness of cabling and physical infrastructure.
- Technology is becoming much less environmentally demanding in terms of power consumption, heat gains, weight and space requirements.

Possibly, the aspect that will have the most significant bearing upon workspace in the short-term, is the inevitable move from CRT monitors to flat panel displays (FPDs), in terms of the amount of area required and its layout. The '20/20 Vision' study of financial trading rooms (Pringle Brandon, 1996), assessed the impact of migrating from CRTs to FPDs as leading to the following.

- **Smaller desks** – the same screen capacity can be achieved in a smaller footprint, resulting in locations of screens and dealer boards no longer being dictated by the size of the monitor, and hence a greater variety of workplace layouts is possible.
- **Closer face to face trading** – the desk depth reduces by up to 36% (i.e. from 1250 to 800 mm), thus reducing the face-to-face dimension from 2500 to 1600 mm, with improved sightlines.
- **Occupant capacity** – the reduced size of the footprint enables more workplaces to be provided within the same overall workspace area – an increase of approximately 24%.
- **PC/processor location** – with a reduction in footprint size there is less scope for CPU locations – where previously workplace size has been dictated by the size of the monitor, increasingly the footprint will be determined by the size of the CPU.
- **Power consumption** – reduced primary and alternative electrical supplies, generator, UPS and switchgear capacities – saving approximately 30% per head, with consequential benefits for overall energy consumption and running costs.
- **Cooling requirement** – reduced heat output leads to reduced cooling loads – saving approximately 20% per head.

- **Display options** – improved front-end visual display options as cross lacing of screens is easier if monitors are driven digitally.
- **Lighting** – highly critical in CRT environments where Category 1 light fittings are required, FPDs with flat, matt finished screens will reduce the incidence of reflections making the use of Category 2 or 3 light fittings possible.
- **Health and Safety** – lighter, smaller screens are easier and safer to install and reconfigure, and with less microwaves emitted, safer to use.

The impact of IT and systems such as the introduction of document image processing (DIP), is having a profound effect upon the way some organizations do their business at present and is likely to have a major impact on much of what today we call 'office work'. Most often the focus in businesses is upon the speed at which the enterprise itself, in terms of the people within it, will accept the introduction of new technologies, with apparently little thought for how such initiatives will be received by customers. Call centres are progressively being introduced to replace the high street branch networks of most banks, in their drive for reductions in the cost of servicing their customers' accounts. But will the loss of the local branch be acceptable to customers? And if so, will customers accept the change at the pace that the banks wish to enact it? And the changes are unlikely to stop there, as banks, insurance companies and other financial services businesses move from a high street branch network to call centres, and then to tele-banking from home via telephone, PC and television links. Each of these new channels to market will have profound effects upon the numbers of people required by the business, their location and the nature of the workspace required, as well as changing the geography of their relationship with their customer – from local, to regional, to national and then to international.

It has long been recognized that an ineffective work environment can have a detrimental effect upon the workers who use it, so why then is it so difficult for many senior managers to comprehend that a well planned and effectively resourced workplace can have positive benefits for those who use it. We find it surprising that many organizations appear not to be able to make the correlation between the quality and effectiveness of the workplaces they provide, and the effectiveness of their personnel and the quality of their output (see Fig. 6.16). For example, most businesses in consumer products industries, freely recognize the value of spending time, effort and money on product design, marketing and advertising because of the effects they will have upon the behaviour of their target customers, i.e. to change them from prospects to buyers. The application of similar thinking to the design and development of the workplace, its promotion and servicing, can have a profound effect upon the behaviour of the people in the business in terms of their operating performance and morale, which in turn can lead to a positive contribution to the performance of their work. Measuring the effect the workplace has upon its users, as referred to in Chapter 3, is a vital part of the facilities manager's role in the business.

6.3.3 Relocation

Where no other satisfactory alternative is identified then the remaining course of action open to the business is to relocate to different premises. In their study (Gibson, 1998) the RICS (Royal Institution of Chartered Surveyors) identified the top three key reasons that drive businesses to make changes to their property portfolios.

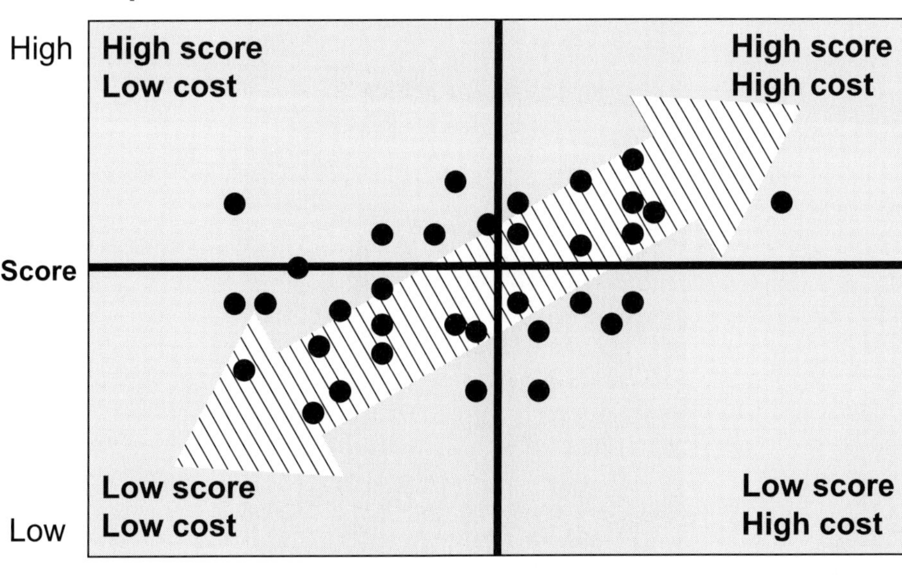

Fig. 6.16 The correlation between facilities provision and customers' satisfaction with their workplaces is of profound importance in effective workspace management (source: AWA Best Practice Benchmarking).

- To reduce cost, quoted by 56% of businesses.
- To dispose of surplus property, 36%.
- To find more appropriate accommodation, i.e. to improve the quality of the work environment, 32% (see Fig. 6.17).

The study also assessed that about 4% of past property reductions were attributable to changes in working practices, i.e. flexible or new styles of working (hotelling and telecommuting), and that this would contribute to about 7% of forecast future reductions. Clearly such initiatives have thus far been seen as low-level reasons for changing buildings. This is all the more surprising as the same businesses considered that internal flexibility and adaptation to cope with business changes, were the most important criteria (almost half rated it their most important criteria), to be applied when selecting new premises.

A USA survey (Deloitte and Touche, 1995) identified a trend for businesses to split work activities into two components, i.e. front office and back office, which was forecast to continue. Further, it was forecast that the back office element would increase in size more rapidly than the front office. This splitting of the business into two finite components, almost inevitably gives rise to the question – is there a need to maintain all of the business in one location? Evidence suggests that when businesses reorganize their workgroups into customer facing front office elements and back offices where the business transactions are processed, this frequently leads to a split in workspace locations. A most graphic example can be seen with the rapid rise of the call centre as a means of conducting business, where many of these activities are accommodated in relatively low cost, and in some cases low grade, accommodation.

Fig. 6.17 The quality of the external environment contributes to attracting businesses from inner city locations to out-of-town locations – Kingston SCL, Edinburgh Park, South Gyle, Edinburgh.

The RICS study also pointed to the limited effect that home-working appeared to have on UK businesses' expectations about the amount of workspace they would require in the future. One third of respondents indicated that the introduction of changing working practices had not given rise to a reduction in the amount of workspace they needed. Given the almost direct correlation between the numbers of people the business requires and the amount of workspace, then at first sight this appears to be an odd conclusion to reach. However, the most likely reason for the apparent contradiction is the comparatively small-scale application of these new work practices and businesses' reluctance to come to quick conclusions about reducing their workspace commitments, which they may subsequently have to reverse. It is dangerous to generalize about the apparent reasons for the slowness to introduce flexible forms of working. The cost of setting up a home-based worker is seen by many businesses as being almost the same as providing office space, which may be used as a reason for delaying or avoiding the adoption of such practices. These situations however fail to take sufficient, if any, account of the intermittent home worker, i.e. a person working *from* home, rather than *at* home. Nor do they appear to appreciate the potential for providing improvements in workspace utilization in central support functions, by employing non-dedicated workplace provisions for personnel who at times may work from home or other remote locations.

When senior UK business managers were asked as part of the RICS study to identify their key expectations for what they considered would contribute most towards a business relocating premises, the following factors emerged, which are ranked in order of the frequency quoted.

- Increased use of IT, quoted by 60% of managers – which is very significant, not only because of the implications for the infrastructure of work environments, but because the effective use of IT is a prerequisite for most the other factors quoted by these managers.

- The adoption of new working practices, 60%.
- A restructuring of the business, 49%.
- Downsizing or reducing the numbers of core employees within the business, 49%.
- Business process re-engineering, 44%.
- Outsourcing non-core activities, 44%.

In addition to influencing some businesses to relocate, we consider that all of these factors could be drivers for change to the design and production of workspace and its use.

The management of a relocation project

The management of a workspace relocation project, be it major or minor, has particular requirements that need to be addressed. The relocation could be a stand-alone exercise or more likely part of a larger project. However, all relocation projects have five identifiable key components.

- **Defined objectives** – a clear unambiguous statement of the business's objectives for the relocation is effectively one of the first elements of the process, as it forms the starting point for the subsequent development of the project, and would typically take the form of a Brief.
- **Time limits** – would include as a minimum, dates for commencement and completion, as well as any interim milestones and dependencies such as expiry of rent free periods, lease notice and termination dates.
- **Prescribed standards** – these can take many forms including performance specifications, traditional specifications, drawings (schematic diagrams, layouts and detailed designs), descriptions, SLAs, instructions and contractual safeguards.
- **Cost limits** – usually established as part of an initial appraisal or feasibility study and then progressively developed throughout the life of the project, e.g. initial estimate, pre-move cost check, tender receipt and report, comparison with estimate and reconciliation, interim cost reports and final account.
- **Resources** – their requirement has to be determined, then appropriate types and quantities of resources obtained and applied to the project in line with a programmed sequence of events; typically they could include labour, plant and materials, fuel and energy, finance, maintenance costs and materials, commissioning and decommissioning, acquisition and disposal.

Workspace relocation projects are temporary activities with clear start and finish points. They therefore require organizations that are of a temporary nature, which brings particular problems for those responsible for managing them. An early step in arranging for the relocation to be carried out is the development of an appropriate project organization. In large scale multi-site businesses there may be a team who are assigned to relocations and move from one to another. However for most businesses a relocation team will need to be assembled for each project. Some relocations can be carried out with a small group formed from within the resources of the facilities management team. Others may require large complex and multidisciplinary teams to be formed where the relocation involves major construction, alterations or changes to services infrastructure and, as such, the appropriate composition of the team will depend upon the specific project circumstances.

Like all projects, relocations need effective management, and experience has shown that this is often best achieved by the appointment of a single person as the relocation project manager. Depending upon the specific project circumstances the project manager may be internal (i.e. an employee within the client organization) or external (i.e. a consultant or contractor engaged by the client for the particular project's duration). However, all successful relocations are the result of the concerted effort of a team of people.

Who should be involved in the relocation project team? The answer will be specific to each project but is likely to include representation from business management and representatives of user departments, IT services, procurement and facilities management, all under the direction of the relocation project manager who of course, in many instances, may be the facilities manager. One of the primary benefits of good project management, is the coordination of activities of all the parties involved in the project and the issuing of clear definitions of responsibility and authority. It is wise to be aware of the potential for sentience; that is, the tendency for an individual to relate to, and to identify with, his own group or discipline at the expense, and to the relative exclusion, of all others. For example members of design teams are often narrowly focused because of their individual specialist skills, which can lead to traditional intransigence, whereas construction teams tend to operate on the principles of diversity and tend to be more 'fluid' in their methods. The relocation project manager has to marshal the efforts and contributions of these quite widely different approaches, and make each appreciate their interdependency and mutual reliance.

Roles and responsibilities

The relocation manager would be responsible for:

- The success of the relocation and usually accountable to a senior executive in the business;
- Control of the other professionals involved in the team;
- Working with functional managers in order to secure the required project resources;
- Coordinating the input of individual specialists within the business, who would be responsible to their respective line managers, as project team members;
- The control and deployment of project resources for the duration of the relocation activities;
- The appointment of all consultants and contractors required to implement the relocation; and
- Where an external agent is appointed he or she would act as the client's representative in all matters delegated to him or her.

The principal activities of the relocation manager would comprise the following.

- **Planning** – involves breaking down the relocation project into analysable components, and working out what time and effort has to go into achieving the completion of each component, then linking the components together in order to work out the sequence and time dependencies between the elements of the project. The objective being to produce a plan of work that shows the dates upon which activities start and finish, and to work out start and finish dates for larger sections of work; determining the relevant resourcing levels required in order to achieve the satisfactory completion of each work component and work section; identifying where resources are over committed and proposing a resource levelling solution; defining required levels of quality and the standards to be achieved in order to carry out effective cost control.

- **Monitoring** – involves observing and recording progress recorded either in graphical form (in the case of time), in tabular form (in the case of costs), in record form (in terms of quality and safety); comparing planned to actual performance for time, expenditure, quality and safety; analysing the impact of any divergences in terms of relative importance; receiving feedback; adjusting the programme in the case of time, budgets (i.e. additional funding or cost cuts) in the case of costs, and improving control in the case of quality and safety; and observing the adjusted system in order to evaluate the degree of improvement brought about by the executed corrective actions.

The key stages of a relocation project

The overall process of management of a project can be said to comprise five stages, namely Initiating, Planning, Organizing and Communicating, Executing and Commissioning, the elements of which are as follows.

- **Initiating** – definition of the relocation and how it relates to the business's accommodation strategy; a statement of scope, nature and objectives for the relocation; a feasibility study to assess the options that may exist (including those to avoid relocation) and their appropriateness for the business; the social and employment context; the legal environment – planning, health and safety and environmental; the commercial environment – trading of business, government incentives, grants etc; the physical environment of the premises and their surroundings; the project strategy – the process, organization, contract procurement and implementation; establishing the success criteria that will be applied to evaluate the effectiveness of the relocation.
- **Planning** – preparation of a brief; defining the scope of work and producing a description of work; estimating costs; programming comprising a master programme and sub-programmes incorporating milestones and control procedures; risk assessment; and procurement strategy.
- **Organizing and communicating** – setting-up appropriate project organization structure; defining responsibilities; establishing a programme for review meetings; communication contacts and procedures; and document issue procedures.
- **Executing** – ensure all parties are performing in relation to their contractual responsibilities and requirements; risk identification and allocation; check insurances; contract procurement procedures, e.g. letters of intent, purchase orders etc; contract administration, including reporting, variation procedures, interim payments etc; monitoring and control – systems whereby cost, time and performance are constantly reviewed against planned levels; implementation of Health and Safety procedures; implementation of construction QA procedures; negotiating and conflict resolution.
- **Commissioning** – planning the commissioning process, which typically will include a return of many specialists to carry out tests independent of others on the project; the hand-over of the 'new' premises to the nominated business representatives; occupation and operation involving fit-out, relocation, training and the putting in place of maintenance procedures; facilities management for the ongoing operation of the building; post occupancy review comprising an audit of the project procedures and an assessment of compliance with the project objectives and user satisfaction.

Risk management

Risks are features of all projects and relocations are no exceptions. The facilities manager therefore needs to adopt effective risk management techniques that should encompass the following.

- **Identification of the risks** – individual, combinations and cumulative effects.
- **Analysis of the risks** – quantify the risks in terms of time, cost, quality and safety, probability and consequences.
- **Ranking of the risks** – categorize in terms of probability, impact, timing and method of protection.
- **Formulating plan(s)** – for each category of risk, and combination of risks – development of a process for avoidance, minimum acceptable level, contingency and disaster recovery.

One of the greatest threats posed by any change programme, is the risk of setting over-ambitious targets. Facilities managers must endeavour to be realistic in their expectations for the speed at which the change can be implemented, the pace at which the forecast benefits will be achieved, and the risks resulting from not achieving the benefits, or achieving them at a slower rate than forecast.

Change control

Often, the biggest challenge faced by the relocation manager is the requirement to implement changes to the premises and their layout during their implementation or construction. The aim must always be to consider both the costs (in terms of time and money) and the benefits (improvements in operational performance) and make a decision on what is in the best interests of the business and its objectives for the relocation (Fig. 6.18).

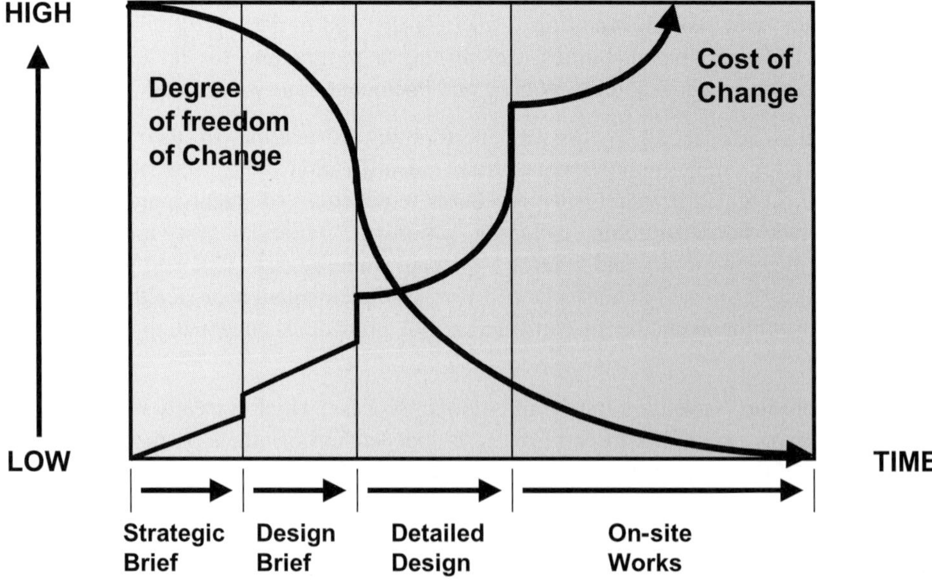

Fig. 6.18 The interaction between the degree of freedom to make change and the cost of change.

The object is to consider any proposed change extremely carefully before it is implemented. Changes can be very expensive once a construction project has started on site, and hence the implications on the programme and final account could be considerable. However, the delay or non-implementation of a change may result in an even greater cost being incurred by the business through later alteration to the completed building, higher maintenance costs and possible disruption to future operations. Change control consists of registering all proposed or potential changes and subjecting them to a detailed investigation before they are actioned. The proposed change would normally be considered in terms of costs to the whole project, the time implications, the effects or value of the change and any other operational effects that will or may result. Each relocation project requires its own change control strategy and operating procedures established at the time of project set-up, which enables the process of robust interrogation of any proposed change to be carried out. The subject of change is one we will develop in Chapter 7.

Ongoing team management

The sound management of the relocation team is an essential prerequisite for a successful relocation project, which would typically embrace the following.

- The briefing of team members, including prior visits to the 'new' premises in order to familiarize themselves with the geography of the premises and how they are, or are intended to be, fitted out.
- Regular progress meetings should be held to cover move-related issues, also calling for reports from other related areas.
- Attendance at meetings should be mandatory; use the 'buddy' system so that if someone is unable to attend then their 'buddy' deputizes, thereby ensuring continuity.
- Be action orientated – do not leave decisions in a void.
- Prompt issue of notes of meetings, listing decisions taken, follow-up action, responsibilities and dates by when actions should be taken or reported.
- Keeping the team well informed throughout the project via a range of communications techniques – briefings, bulletins and reports.
- Selection of contractors based upon pre-qualification and tender, including Health and Safety procedures, labour relations, management, pre-planning, phasing, duration and hours of working, levels of insurance cover and after-move service.
- Moving instructions should be issued to all affected by the relocation, and should include programme dates and times; roles and responsibilities; destruction of old material – rubbish disposal before, during and after the move; requirements for crates and specialist equipment; labelling arrangements and coding; what is to happen to items that are not to be moved; arrangements for access to the new premises; communications systems during and after the move; new rules to apply and help desk arrangements.

Contingency plans will need to be developed to cover the many aspects of the relocation, and typically would include such issues as the following.

- Failure of the lifts in either or both of the old and new buildings – should a lift engineer be in attendance?

- The inability to gain vehicle access to either or both of the old and new buildings – are alternative routes and accesses possible?
- If progress falls behind programme – can more resources be made available? Or can the team work longer hours?
- Services failure – electricity, telecomms, WANs, air conditioning, heating, lights – make provisional arrangements for emergency hiring.
- The outbreak of fire in both the old and new buildings – impose smoking restrictions.
- Major damage to items being transported, and to the old and new buildings.
- Failure of the vehicle(s) used for the relocation – their damage, loss or theft.

Once the relocation has been completed, the facilities manager and the project team, should conduct a post occupancy evaluation, which would seek to establish the following.

- The effectiveness of meeting the declared objectives of the relocation – were targets and objectives met? Are the original objectives still valid? What 'refinements' need to be carried out?
- User satisfaction – before and after appraisal and be sure to follow-up the results.
- Audit the project processes and the effectiveness of the procedures followed, after which the relevant feedback should be captured with a view to applying the experiences gained for the benefit of future projects.

6.4 Flexible workspace

What we have grown to call office work is central to all developed societies. Arguably, over half of those in employment now work in what we have learned to call 'office space', with the numbers increasing inexorably as we move from manufacturing increasingly towards knowledge-based work. Over the last century work has changed enormously – in quality as well as quantity. Yet the basic rules for locating, designing, planning and constructing workspace have not really changed since the beginning of the century. These long lasting rules stem from Frederick Taylor's globally influential notions of 'Scientific management', which depend fundamentally upon treating employees not as self-directed individuals but as units of production. This may help to explain why most office environments do not, of themselves, generate much pleasure or enjoyment for their users.

Today, organizations need to be more responsive and innovative than they have been in the past, in order to develop competitive advantage. This places even greater strains upon the business's workspace. The location as well as the pattern of work has changed and will continue to evolve in response to competitive business pressures that will be experienced by all enterprises.

Traditional organizational models and workplace environments were never designed to cope with rapid market changes because they have their roots in the notions of hierarchy and permanence. The changing patterns of work renders such concepts obsolete, as businesses are constantly seeking ways to re-invent themselves. So organizations must be able to change their shape and mode of working with the minimum amount of disruption to their day-to-day operations and without incurring financial penalties.

6.4.1 Flexible working concepts

Flexible working is a term that has been used to describe changing work patterns where people work at the most appropriate time and place to meet the needs of the business and of the individual (Fig. 6.19). It may embrace the notion of home-based working, where people spend some time working from their home. It may embrace teleworking, where people work from remote centres that are 'connected' to the central 'hub' of the business by electronic means. It may embrace the idea of 'hot desking', where people when they are in the office, use a workplace that is time-shared with others. For more structured roles it may embrace 'hotelling' which is a more disciplined style of hot desking in which the workplace is readied for the arrival of a worker who has booked the space in advance. This often includes the setting up of the workplace with the individual's storage and personal effects such as pictures of the family and personal artefacts.

Many of the flexible styles of working are based upon the generic concept of *address free* workplaces. This concept is founded upon the principle that staff who habitually work out of the office building, do not require a dedicated workplace. Consequently, when working within the office, these itinerant workers make use of workplaces of a standard configuration allocated for this purpose on a first-come first-served basis. Each person working in an address free environment will have their own mobile storage unit, a pedestal

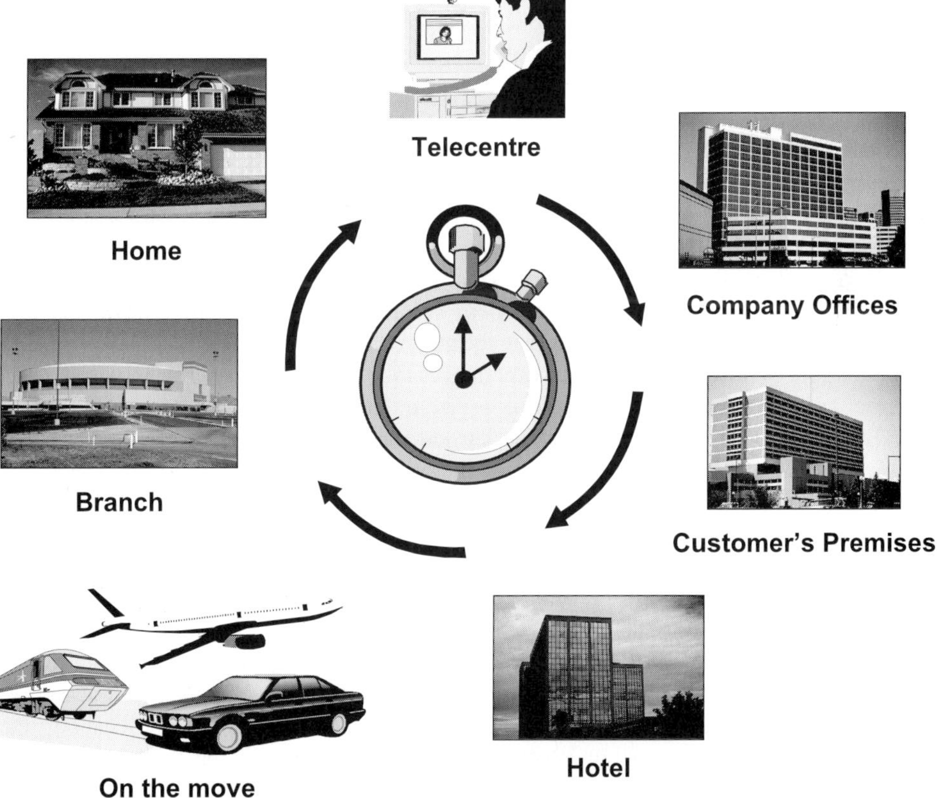

Fig. 6.19 Flexible styles of working enable people to work in a range of locations.

unit on castors or a pilot's case, in which they can store their work materials, personal files and belongings. Upon arriving at a workplace the user is able to log-on to the network advising that they are now 'resident' at this particular location, which then permits them to gain access to relevant data systems. The act of logging on to the network also activates the telephone system such that incoming calls for the user in residence are routed to them at their predetermined extension number. This kind of arrangement is only possible where workplaces are connected via networks (LAN and WAN), and are provided with a common IT platform, operating system and applications.

The introduction of so-called flexible styles of working such as hot desking, was aimed at addressing the needs of itinerant workers. If it is considered appropriate to permit people who use the workspace as a transit location, to choose work settings that best suit their needs, then why not extend the same facility to those who are 'permanently' based in the premises – the concept we call *Right Working*.

Because of increasing competitive business pressures coupled with an increased reliance on the work of the individual through empowerment, the workplace needs to be tailored to the functional needs of the individual and their work processes as never before. The 'traditional' approach of providing one dedicated workplace for every employee is failing to address true operational needs. This is because the majority of people today carry out a wide range of multitasking activities, rather than older, more conventional styles of working that were largely based upon mono-tasking, i.e. carrying out a single, simple repetitious activity. To be effective, multitasking workers require settings that are appropriate to the needs of each component task. This requirement cannot be addressed by a conventional approach that is founded upon providing a dedicated workplace to each person, as it ends up being a compromise in meeting the different and often conflicting, functional requirements of specific tasks.

Instead, the *Right Working* approach is based upon providing each individual with a range of taskplaces at which they can choose to work, to meet the specific demands of a task at any point in time such as quietness, group interaction, confidentiality, telephone communications, etc (Fig. 6.20). Consequently, throughout the working day, week, month, people move from workplace to workplace as their varying needs require.

Why work at a single workplace? The convention of providing a single workplace that is dedicated to an individual has always seemed a strange concept, particularly since at school, college and university, our experiences of work environments is based upon moving from classroom to classroom, or from lecture theatre to library, i.e. from workplace to workplace. When IBM were introducing flexible working practices into their UK facilities, they needed to communicate the basis of this fundamentally different way of working. As Peter Wingrave, former Workplace and Design Strategy Manager at IBM UK, explains it, 'IBM adopted the slogan to move from *mine to have* to *ours to share*', which beautifully encapsulates what flexible styles of working is really all about: that is, equipping the organization to best meet the needs of business that is increasingly being carried out by teams rather than individuals.

The principal objectives of *Right Working* are to provide a range of environments and facilities that are aimed at supporting people (individually and collectively) in achieving optimum performance in their work, improving their quality of life, while enabling the business to optimize the use of its resources. An imbalance in the three will almost inevitably lead to dissatisfaction. One advantage the advent of flexible styles of working will bring is, according to Charles Handy and others, the demise of archaic functional

Fig. 6.20 Right working provides a range of different task settings available to all workers on a shared basis.

silos, so long a characteristic of most businesses, to be replaced by process-focused, cross-functional work groups.

Today, although there are clear signs that the 'winds of change' are blowing through businesses, power and politics in the workplace are still evident. In many organizations, rank and status still guide the allocation of workspace, which in its most extravagant form is manifest in the provision of the dedicated, single occupancy enclosed office. While there are likely to be instances where the work needs of the individual warrant such a provision, in many businesses the enclosed workplace is seen by many as a 'badge of rank' and a meeting table as denoting a 'three star general'. Tools, for that is what workplaces are, should be provided and available to those who need them to meet the objectives of the business productively.

Where flexible styles of working are proposed, which involve sharing desks, Health and Safety considerations need to be reviewed to ensure that each user has the capability to adjust the workplace to suit their individual needs. When observing people in cars, it is interesting to note how many drivers adjust their seat position (forward or back, height, and angle of rake) and also the mirror positions. Although routines such as these are second nature to motorists, they have not yet penetrated the workplace.

A recent study states that open-plan workplaces cannot be effective for work because they are too noisy with too many distractions. But if we look at a reading room in any collegiate library we will see many scholars working very effectively in an environment where one can hear a pin drop. Why? Because the workplace setting and the protocols which guide its use, are predicated on the creation of an environment that truly supports the needs of its users. It is not necessary to erect walls, to avoid distraction, just eliminate the cause of the distraction. So within a reading room, telephones of all types (mobile or otherwise) are prohibited, speech is only permitted in hushed tones, and if people want to converse and exchange ideas then separate meeting rooms are provided for those purposes.

Protocols are simply the forms of 'etiquette' observed by members of an organization

(a)

(b)

Fig. 6.21(a) and (b) Flexible working environment at Scottish Enterprise, Glasgow, Workplace of the Future.

that define the social 'norm' for that community. As the work community within the business requires collaboration and the sharing of space and facilities, it is important for everyone to understand the protocols that will enhance their interaction, support privacy needs and promote effective utilization of resources. Below, we list some examples of protocols that businesses could adopt when implementing shared workplace regimes.

- **Booking procedures** – where workplaces, rooms and equipment are available for advanced booking (in periods of 15 minutes), users should advise if they no longer require the facility and also when finishing early, thereby making it available for others.

- **Non-bookable spaces** – are generally intended for continuous use for up to a maximum of 2 hours.
- **Furniture and equipment configurations** – when finishing with the use of a facility, furniture and equipment should be returned to its original layout and settings, making it ready for the next user.
- **Presentation boards** – after use, whiteboards should be erased and used flipchart pages should be removed, with the space left ready for the next user.
- **Privacy at the workplace** – can be signalled by placing 'do not disturb' signs at the workplace, which should be observed by everyone.
- **Self clear** – cups, plates etc used at any setting should be cleared and surfaces left clean, this includes the removal of all papers, including waste, making it ready for the next user.
- **Enclosed quiet workplaces** – when the door is closed the occupant is signalling his/her desire not to be disturbed by others outside the room, unless they have previously advised an alternative procedure to be adopted.
- **Open-plan quiet workplaces** – those using open areas designated for quiet working, should respect the work needs of others by not using mobile communications or holding meetings in these areas and by moderating their speech levels so as not to cause disturbance.

Through more intensive and prolonged periods of use, flexible working arrangements can put added stress and pressures on facilities services and their maintenance. This will feature in shorter life expectancies of products and materials, reduced time-windows in which to conduct repairs and maintenance operations, as well as requiring the addition of services that may not have previously been required. New styles of working also require new types of facilities services. The provision of space and facilities booking services similar to those used in a hotel or club, stewarding services to support the increased intensity of use, and the management of shared information, are all likely to feature in the facilities manager's expanding array of services.

As businesses adopt different styles of working, they need to reappraise the way in which they review their occupancy costs. No longer will it be sufficient to consider these costs in terms of square metre rates, as this measure is inappropriate for the varied way in which the space is used. Occupancy costs need therefore to be assessed in terms of rates per workplace, and even more importantly, per person. In *New Environments for Working* Laing *et al.* (1998, p. 40), indicate how the costs of creating different types of workplace settings – cell, den, hive and club (see next section for definitions) – can be put into an appropriate context when taking account of the typical occupant densities that each can achieve – cell 22.2 sqm/person, den 14.7 sqm/person, hive 10.4 sqm/person and club 7.6 sqm/person. On this basis, Laing advises that overall workspace life cycle costs per head rise with work patterns – from club, to hive, to den and finally to cell.

6.4.2 Workplace settings

As we have explained, to meet the flexible working needs of the organization, a range of workplace settings is required, each of which comprises four elements.

- **Physical environment** – the provision of appropriate workplace settings and hardware.
- **Information environment** – the availability of information and ready access to shared knowledge, in hardcopy and electronic form as well as direct from people and via communications systems.
- **Service environment** – the provision of support, servicing and maintenance for the workspace and its infrastructure to sustain it and its users at optimum operating performance.
- **Protocol environment** – the principles and procedures that guide the use and management of workspace.

Effective workplaces will be those that are based upon a holistic approach to their design, which incorporates a balance of these four environments. 'Awareness of how office buildings are managed is critical', says Frank Duffy (1998, p. 70), 'because even the most sophisticated and superb office environments are useless unless they can be made to work properly at all scales and through time.' Duffy describes four workplace settings, which he calls the hive, cell, den and club, that demonstrate how these attributes can be combined to meet different work needs.

- **Hives** – characterized by individual, routine-process work with low levels of interaction and autonomy. Hive workers sit at simple workplace settings for long periods of time, usually on a nine-to-five schedule. Variants of hive offices include 24-hour shift working. Workplace settings are typically open-plan, screened and impersonal. Typical work groups include telesales, data processing, financial and administrative operations and basic information services.
- **Cells** – accommodate individual, concentrated work with little interaction. Highly autonomous people occupy these fully enclosed offices, on an intermittent and irregular pattern, with extended working days, often working from other locations such as home or clients' premises or from transient locations such as hotels. Each individual workplace must be designed to provide for a complex variety of tasks. The autonomous pattern of work, implying a sporadic and irregular occupancy, means that the potential exists for such work settings to be shared. Typical occupiers of cellular offices include accountants, lawyers, management and employment consultants, and computer scientists.
- **Dens** – are associated with group work, typically highly interactive but not necessarily highly autonomous. As they are designed to support group work they often comprise a range of simple settings, usually arranged in open-plan workspace or in group rooms. While the settings are normally designed on the assumption that individual workers occupy their 'own' desks, such groups are also likely to need access to local ancillary space for meetings and project work, and for shared equipment such as printers, copiers and other technical facilities. Although tasks are often short-term and intense, sometimes they may be more long-term, and always involve collaborative working. Typical work requiring dens includes design, insurance processing, some media work, particularly radio and television, and advertising.
- **Clubs** – are for knowledge work; that is, work that transcends data-handling because it can only be done through exercising considerable judgement and intelligence.

Typically, such work is both highly autonomous and highly interactive. The pattern of occupancy tends to be intermittent over an extended working day. A wide variety of time-shared and task-based settings serve both concentrated individual and group interactive work. Individuals and teams occupy space on an 'as-needed' basis, moving around it to take advantage of a wide range of facilities. The ratio of sharing depends on the precise content of the work activity and the mix of in-house and out-of-office working, possibly combining teleworking, homeworking, and working from client and other locations. Typically, the work carried out in clubs would encompass open-ended problem solving, by highly intellectual people, who need easy access to an array of shared knowledge, and would include creative teams such as advertising, marketing and media groups, IT groups, and management consultancies.

While hives, cells, dens and clubs can be seen as characterizations of whole businesses, we consider that, in practice, there are many organizations that may require elements of each type of workplace to address the needs of different work groups. For example, the legal department of a financial services business is likely to have no less of a need for cell environments than would a legal practice, but will also need hives for processing staff, clubs for marketing staff and dens for product development teams. In short, most business will have a requirement for aspects of each type of environment, and these requirements are likely to change as the business evolves its operations.

Case Note 6b

Healthcare insurance company, Southern England. As part of its introduction of flexible styles of working, this internal consultancy of approximately 20 people in an insurance company redefined the work environment it required. Instead of the conventional open-plan and enclosed cells that had previously been provided, and were in use throughout the rest of the business, this work group sought an environment that was more responsive to their itinerant work patterns and their own idiosyncrasies. When they were in their group's area it was usually to work with colleagues, exchange ideas, hold meetings or conduct research from their reference materials. To address their needs, a range of work settings were provided that incorporated facilities for quiet working, touch-down spaces for telephone and e-mail activities, and meetings spaces of different types, such as formal meetings seated around a table, high level counters for stand-up meetings and soft seating areas (incorporating bean bags, fish tank and even a fountain) for informal and social meetings. Only one conventional workplace was provided for use by an administrator, who was the only member of the team based permanently in the office.

Case Note 6c

Financial services company, Southern England. A pilot flexible working programme that was being designed for a work group in this financial services company, incorporated a soft seating area for use for informal meetings and relaxing. When the designs were tabled staff commented that the location of the soft seating area near the access to the work group would cause them difficulties. Upon investigation it was found that the proposed location could be overlooked by people passing the work group's area, which 'might give the wrong impression' that staff were not working but relaxing and enjoying themselves. Heaven forbid that people have fun in the workplace! Such examples speak volumes to us about the attitudes of the staff, and the management regime that prevails in the organization.

References

Akhlaghi, F. (1993) The business of space. In: Paper presented at a conference on *Quality Concepts and Premises Management in Further Education*, Sheffield Hallam University, 30 April.

Becker, F. and Steele, F. (1995) *Workspace by Design: Mapping the High-Performance Workscape* (San Francisco: Jossey-Bass).

Bon, R. (1989) *Building as an Economic Process* (Englewood Cliffs, New Jersey: Prentice Hall).

Deloitte and Touche (1995) *Space Needs of Corporate Headquarters in America* (Chicago: Deloitte and Touche Real Estate Service).

Duffy, F. (1998) *The New Office* (London: Conran Octopus).

Gallup (1996) *Shaping the Workplace for Profit*. Commissioned by Workplace Management.

Gibson, V. (1998) Changing business practice and its impact on occupational property portfolios. *Right Space: Right Price?* **4** (London: The Royal Institution of Chartered Surveyors).

Handy, C. (1997) Boring workplace, boring worker. *Management Today* November, p. 29.

Joroff, M., Louargand, M., Lambert, S. and Becker, F. (1993) *Strategic Management of the Fifth Resource: Corporate Real Estate. Report of Phase One – Corporate Real Estate 2000*. The Industrial Development Research Foundation (IDRF).

Laing, A., Duffy, F., Jaunzens, D. and Willis, S. (1998) *New Environments for Working: The Re-design of Offices and Environmental Systems for New Ways of Working* (London: E&FN Spon).

McGregor, W.R. (1994) Designing a 'Learning Building'. *Facilities* **12** (3), 9–13.

Nourse, H.O. and Roulac, S.E. (1993) Linking real estate decisions to corporate strategy. *The Journal of Real Estate Research* **8** (4), 492.

Pringle Brandon (1996) *20/20 Vision: A Report on the Impact of Flat Panel Displays on Trading Room Design* (London: Pringle Brandon).

Propst, R. (1968) *The Office – A Facility Based on Change* (Michigan, USA: Herman Miller).

The Henley Centre (1996) *The Milliken Report: Space Futures*. Commissioned by Milliken Carpets.

Then, D.S.S. (1996) *A Study of Organisational Response to the Management of Operational Property Assets and Facilities Support Services as a Business Resource – Real Estate Asset Management*. Unpublished Thesis. Heriot-Watt University, Edinburgh, Scotland.

7

Managing space demand over time

In the previous chapter we saw how the need to reconcile demand for, and supply of, appropriate workspace is, for many organizations, as much about corporate survival as it is about developing their competitive edge by managing business resources more effectively. Technology and increased flexibility in the workplace has raised questions about the efficient use of workspace, especially for office space. In this chapter we consider the issues of managing space demand as a business resource, over time.

As we pass through the 'trans-millennium age', it is becoming abundantly clear to public and private organizations alike, that the notion of *time* is synonymous with *change*. The pace of change is, in many ways, already starting to redefine our lifestyles, the way we work, where and when we work and how we interact. A vision of this journey of change for organizations is summarized by Laing *et al.* (1998, p. 15), as comprising:

- How we work

From	**To**
Routine processes	Creative knowledge work
Individual tasks	Groups, teams and projects
Alone	Interactive

- Where we work

From	**To**
Places	Networks
Central	Dispersed
Transport	Communication
Fixed location	Multiple locations including the home

- The use of information technology

From	**To**
Data	Knowledge
Central	Distributed
Mainframe	PC, video, telecomm, e-mail and Internet
One fixed interface	Mobile, personal, nomadic and virtual
Big – desktop	Small – palmtop, pocket, laptop

- Using space over time:

From	**To**
One desk one person	Shared group and individual settings
Hierarchical space standards	Diverse task-based space
9 to 5 at one place	Anywhere, anytime
Under-utilization	Varied patterns of high intensity use
Owned	Shared

In order for facilities managers to contribute fully to the productive processes of the organizations they are supporting, they must have a thorough understanding of the influences driving the above changes, and hence the means to facilitate and to manage the change process.

7.1 Managing change

Organizations seldom remain static. The political, economic and technical climates surrounding them change almost constantly, affecting the markets they work in, the products they deal in, the technology available to do their jobs and the activities they undertake. As an organization faces change, so too does its workplaces, as they are subjected to many and diverse pressures, most if not all demanding change. Convergence of industrial and commercial working practices, together with an ever-increasing sophistication in people's demands, make change inevitable and the control and management of that change essential. Under these prevailing conditions, the most effective organizations are those who face up to the reality of change and plan, implement and control change, to their advantage. Nonetheless, the human animal appears to have a reaction against change. Fear of change, i.e. a natural resistance to change, emanates – we believe – from fear of the unknown and discomfort at the pace of change, where the individual often prefers to retain the status quo despite subscribing to its shortcomings, rather than embracing change, the outcome of which is unknown.

Change in all its guises is probably the most significant component of a facilities manager's job: most significant in terms of the time and effort it will absorb – most significant in terms of the diversity and complexity of the factors involved – and most significant of all, in terms of the impact it can have upon the business and its effectiveness. Consequently, the ability to manage change successfully must rank as a principal attribute of every practising facilities manager.

7.1.1 What is change?

Almost all of our consultancy work involves the processes of change and dealing with the responses of people facing change. Frequently we hear people we encounter through our work, in their attempts to justify a change to a disbelieving colleague, describe it as 'a necessary evil'. The use of this phrase by those championing the cause of change, was they believed appropriate, as it expressed a supportive view of a changing situation bringing out the positive attribute of something unpleasant. However, behind the use of this phrase lies a mental barrier, created by negative thinking. It stems from the belief that there is something wrong and to be avoided in changes, all changes. The phrase is not only inappropriate but can be self-defeating, as some changes are certainly not necessary, and many are quite definitely not evil. Despite apparent reticence to change, which all facilities managers will have encountered, most supported by incredibly ingenious and often convoluted arguments to justify maintenance of the status quo, we all accept change in our daily lives and many of us go to great lengths to inflict change upon ourselves. This is evidenced by the dramatic growth in DIY where change is self-inflicted, often with apparent masochistic delight. So too do we respond to the fickle changes of 'designer' fashion, which so many want to follow in its every turn. And, arguably the most demanding change which most of us encounter each year, the great exodus in the summer months to change the cultural and climatic environment in which we live, for a brief two weeks.

With all this self-inflicted change in our daily lives, why then does accepting change in our work environment present such a problem? Primarily the 'problem' emanates from situations of forced change, which we define as situations where: the recipients of the change are not the initiators of the change, or even worse, where the initiators are not themselves the recipients.

As a first step to managing change, it is important to recognize and understand the different types of change that exist, their causes, implications for the organization and, as a result, the tactics that are necessary to employ to bring about successful implementation. At its most basic level, all changes are either additions or subtractions. Even exchanges are a combination of an addition following a subtraction. This applies to all situations irrespective of their scale. From adding a PC to a workplace, to removing a partition from a room – all are either additions or subtractions. Even relocating premises can be seen as an exchange, comprising the subtraction of the old building and the adding of the new.

From our early school days, we are taught that additions are positive and subtractions are negative.

+ Addition	– Subtraction
Positive	Negative
Good	Bad

Our very use of such symbols accentuates these groupings and consequently our thinking. Is it therefore surprising that we find it difficult to comprehend how a subtraction can be of positive benefit, and an addition of negative value? It is here that the 'necessary evil' concept first takes root. However, there can be little doubt for example, that the addition of high brightness lighting into an environment to be used by VDU operators, is anything other than a negative or bad practice. From this, and many other similar examples, our evaluation of an addition or subtraction can be said to include a vital element, namely appropriateness. In other words, to determine whether a change (addition or subtraction) is beneficial, one needs to examine more than just the content of the change itself, but also its effects on, and implications for, other aspects of the business's operations and its people.

A key factor therefore in successfully presenting proposals for change for acceptance by others, is an understanding that not all additions are good, and equally that not all subtractions are bad. It also needs to be appreciated that the benefits of the change may be, and often are, different for the business than for the people in the business. Once the facilities manager understands these issues, it makes the task of communicating change to others, a great deal easier.

7.1.2 Communicating change

It could be said that 'one man's proactive change is another's reactive change'. What this is meant to convey is that, as part of the process of managing change, *communication* is vital. If a change is planned in advance, but is not communicated to those who will be affected by the change, then when it is time to implement the change those affected will be reacting to a trigger. This is true for those affected either by the change itself, or by the process of implementing the change. In both cases lack of communication will lead to lack of understanding, which will in turn lead to the creation of defensive barriers.

Of course it is neither practicable nor desirable to advise everyone of all planned changes, since to do so may have the detrimental effect of distracting them from the tasks in hand, or possibly worse by demotivating them until the changes have been implemented. The timescales between planning the change, communicating the details of the change and implementing the change are crucial to its success. Therefore, for efficient and expeditious management of change, it is essential to communicate, educate and inform all of those affected by the change.

People are not machines. Recognizing this distinction is important, as although people can be directed, they must also be treated with respect as, unlike machines, a person's output is governed by the way they 'want' to respond, and not solely by the way they 'can' respond. Consequently, the skills required for successful management of change, particularly with regard to its impact on personnel, have their roots in human resource management. Never forget that changing facilities, for good or bad, will always impact on people. As Barrett (1995, p. 177) advises 'putting the human element into change' is vital for success. A key aspect of this human face of change, is ensuring that the communication process involves the articulation of the components of change and the explanation of the benefits for all of those affected by it, both directly and indirectly. Such explanations should be for the individuals in the business and not just for the organization as a whole. It should show how individuals and their teams can become more efficient, effective and productive, and how this in turn will be reflected in the improved performance of the organization, which leads to benefits (financial and spiritual) for the individual.

To be at its most effective, the communication process needs to start at the earliest possible stage, and will involve the facilities manager in consciously *taking people with him or her* through the journey that is change. Early consultation is advocated not only to advise the nature and type of change planned, but hopefully also to elicit from individuals their contribution to the justification of the need for change, as well as their influence on its requirements, process and implementation. Through engaging those affected by the change in the process from its earliest stages, it makes them feel *part of* the process and not *apart from* the process. The aim being, where practicable, not only to have them subscribe to the need for change but hopefully also to gain benefits from their active participation. Experience has frequently shown that people do not resist change if:

- It benefits them – the distinction between personal and corporate benefits needs to be well defined;
- They are allowed to participate – rather than being forced into a change, a successful outcome is far more likely to be achieved where people are allowed to play an active part in the process; and

- They have some control over the outcome – developing from their participation, involvement in the decision making process often leads to 'ownership' of the result.

7.1.3 Planning for change

Forewarned is forearmed, and this was never more relevant than to the implementation of workspace change. As was said in Chapter 3, we cannot predict the future but we can prepare for it. Much of the 'business of space' is about assessing the future needs of the organization and then developing a plan to address those needs. But how do we develop a plan that can remain valid for anything but the shortest of timescales? As we mentioned earlier, one of the principal objectives of the development of an accommodation strategy is the creation of a plan for how the workspace needs of the business can be addressed through time. Whatever process is adopted for the development of an accommodation plan, any plan for that matter, the facilities manager should found his or her plan upon three basic tenets.

- **Plan for what you can foresee** – the obvious and often apparently the only basis for planning, focuses upon the identified needs of the business.
- **Plan for what you cannot foresee** – the unexpected or at least the unpredictable in terms of timing and extent, or both, which will include business continuity planning as well as disaster recovery measures.
- **Plan for what you can reasonably expect to happen** – based upon trends, extrapolations of, and projections from, information about what can be foreseen, which makes it more judgemental than the other two elements.

Foretelling the future is made considerably easier if you have access to the plans upon which the future will be based. However, with the best will in the world you cannot realistically expect to predict the future with any degree of accuracy, but you can prepare for it. The facilities plan must be integrated with the corporate or business plans, to let the facilities manager see the overall objectives of the organization. Linking the planning process with the day-to-day operational issues of business needs is what facilities management is all about. So too is the management of change. Success in the ability to respond to change, is largely governed by the success of forward planning, both in forecasting the timing of likely change as well as the type and amount of resources required.

In this way, a proactive stance to change can be taken. How much accommodation will be required over the planning periods of 2, 5, 10 and 25 years? What type of premises and infrastructure will be required to meet the future business needs? From the answers to questions such as these it will be possible to develop a plan that will identify, through time, the amount, type and location of premises required. This will then permit an investigation to be carried out to determine the sources of the required provision (conversion, refurbishment, or new) if they in fact exist. It may be that by planning ahead for change, an easier, less painful route can be found than might otherwise have been followed. In some cases, for example bearing the pain earlier by relocating, may permit future growth requirements to be accommodated, whereas to remain in the existing premises would be too great a restriction in the future. This type of forward planning avoids the situation of investing in facilities that would have to be written-off over the short term, when relocation is finally forced upon the organization as inevitable.

Disaster planning and the handling of resultant changes, is a detailed subject in its own right and forms an important part of an organization's risk management programme. Consequently, it is neither feasible nor appropriate to consider this subject in much depth here. Suffice it to say however, that in developing an organization's strategy for the management of change, those strategies resulting from disasters and emergencies should not be overlooked.

7.1.4 Organizing for change

The size of the change team, its technical skills and its management structure will be different for every organization. However, it is important to recognize that those involved with the implementation of change require different skills and techniques from those responsible for installing a facility in the first place. A clear example would be the separation of different teams for routine maintenance work from those responsible for implementing reactive changes. This is particularly noticeable in organizations that experience high levels of churn. The motivation of those responsible for planned change as opposed to reactive change is different. In the first, the individuals involved have the benefits of forward planning in that they have longer lead times available to them, for materials and services, and the pattern of work tends to be more routine and repetitive in nature. Those responsible for reactive change however, often have to do so with the minimum of notice, in which instances long lead times are replaced by holding stocks, and the routine is replaced by the one-off. Of course, it is possible for the same individuals to be involved in implementing proactive and reactive change, however, the different attributes of both beg a comparison similar to that between the marathon runner and the sprinter. Both runners are in prime physical condition, and committed to giving of their best in their pursuit of the prize; however, neither is ideally suited to tackling the task of the other.

There are often good reasons why a business will engage the services of a consultant to manage the process of change – lack of appropriate in-house resources being the most prevalent. It is important in such circumstances that the managers of the business spare no efforts in communicating the reasons for engaging a third party, so as to avoid widespread misconceptions of the 'business disowning' or 'walking away from' the change and all it represents. Failure to do so effectively may compromise the intentions of the change, frustrate the efforts of those involved and ultimately threaten the change process itself.

7.1.5 The Process of change

The process of implementing all changes has eight discrete stages (see Fig. 7.1) comprising the following.

- **Start** – either as a planned and programmed event, or as a result of a trigger that initiated the change or a request for the change.
- **Evaluation** – assessment of the change, its objectives and what it entails; the benefits and risks to the business and its people that could emanate from its implementation or rejection.

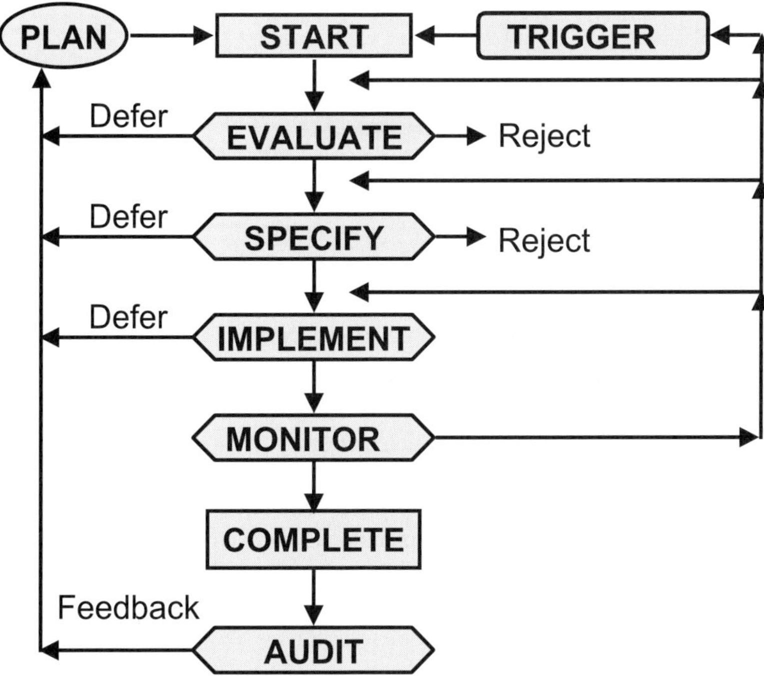

Fig. 7.1 The key stages of the process of change.

- **Specification** – detailed definition of the change, its prerequisites, the resources required for its implementation, a programme of activities and a definition of the success criteria by which the effect of the change will be assessed.
- **Implementation** – the means by which the change is enacted.
- **Monitoring** – the ongoing process of reviewing the progress of the implementation of the change and its compliance with procedures, standards and programme.
- **Completion** – the point at which the implementation of the change finishes, the new way comes into operation and the old way is destroyed.
- **Audit** – the process of assessing the effectiveness of the change in terms of achievement of the desired objectives, and the effectiveness of the procedures employed.
- **Feedback** – the operation of retaining knowledge and experiences gained from all stages of the change process, which will involve updating procedures, records and the process itself, for the benefit of future change initiatives.

There are many different ways of implementing changes through time, which we consider can be derived from six basic forms as illustrated below (see Figs 7.2 and 7.3).

Each form of change has its own application, the key characteristics of which are as follows.

- **Ballistic** – with a rapid climb of the 'learning curve' and gradual decreasing of the rate of change.

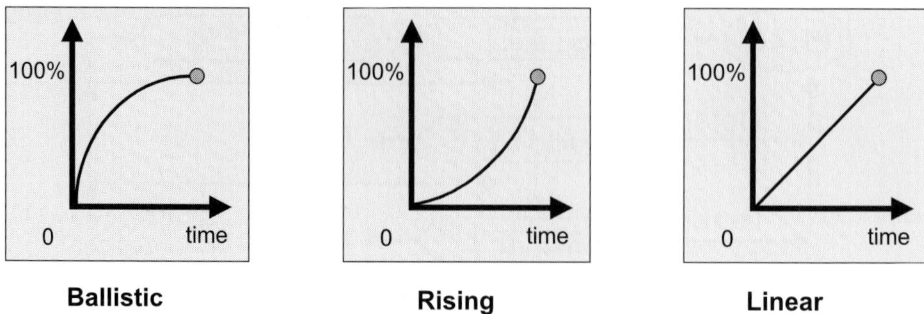

Fig. 7.2 Illustrations of ballistic change, rising change and linear change.

- **Rising** – the mirror image of ballistic change, rising change is typified by a gradual start with the rate of change progressively increasing.
- **Linear** – where the process of change is constant from start to finish.
- **Instant** – where all of the change is implemented at one time.
- **Stepped** – in which the change is implemented in discrete stages of either equal or different increments.
- **Erratic** – where the changes are carried out in an ad hoc manner, most commonly associated with a trial and error approach.

For all change, no matter its type, there are six essential prerequisites for its effective management.

- **Sponsor** – the person who acts as the *champion* of the change, and as its leader, should be both knowledgeable about the change and experienced in dealing with the issues that are likely to arise en route; on major change initiatives the sponsor is likely to be a senior executive responsible to the business's Board or senior management team for the effectiveness of the change.

- **Strategy** – a key component of facilities and accommodation strategies is a strategic plan for making changes, which should be linked to the business plan of the organization.

- **Consultation** – a process for involving the customers and beneficiaries of the change as well as all those affected by the change, seeking their input to the requirements of the change, its timing and the process for implementation.

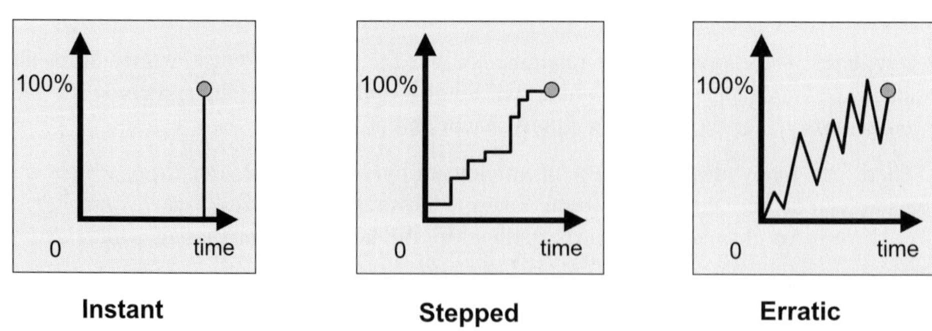

Fig. 7.3 Illustrations of instant change, stepped change and erratic change.

- **Communication** – an effective means of two-way communication between those charged with designing and implementing the change, and those who will be affected by it, keeping each other informed of progress and developments.
- **Planning** – the process by which the change will be specified, designed, procured, implemented and evaluated.
- **Evaluation** – the criteria and method to be employed to measure the effectiveness of the change and its success in delivering the desired objectives to the business.

To be effective, all changes need to be owned. During its formative stages, the sponsor and the change management team will own the change. Progressively, as the structure of the proposed change is developed, and its components are designed and implemented, then *ownership* of the change must be transferred from those involved in its implementation to those who will make use of it and be responsible for its operation. An important aspect of managing the change process, which is often overlooked, is what is to happen to the old way, i.e. the situation that prevailed before the change was implemented. All too frequently, the team responsible for implementing the change focus upon the construction of the new way and tend to ignore the old way, as if expecting it to wither on the vine. However, where the old way is left to take care of itself then, when problems arise with the new way, people will revert to their old comfortable friend, i.e. the former method of working, making the introduction of the change more difficult than it might otherwise have been. Resolution of the future for the old way therefore needs to be addressed every bit as consciously as the intentions for the new way. Every change strategy should therefore include deliberate plans for the removal or the destruction of the old way as an agent for successful implementation of the proposed change.

A further trap that always lies in wait for the unsuspecting facilities manager, is the *transition* phase – that is, the period during which both the old and the new ways exist alongside each other. Clearly there are some changes where this would neither be practical nor desirable – changing from driving on the left-hand side of the road to driving on the right, is one situation where a transition phase should be avoided! We believe that wherever possible all changes to the work environment should be implemented without transition phases, as experience has shown that they can often cause more problems than they solve, while delaying if not impeding the benefits of the new way from being achieved. Wherever possible, a pilot programme should be employed to test the principles of the change, prior to its roll-out on a larger scale. Pilot testing can bring benefits in terms of improved understanding of the effects the changes will have and, in so doing, help in avoiding the need for a transition phase. Where transition phases are unavoidable then they should be governed by the following principles.

- Set deadlines for their start and finish.
- Make their duration as brief as possible.
- Adhere to the time-scales, otherwise pressures for extension after extension are likely to ensue.

7.1.6 Measuring the effectiveness of change

In his book *Total Quality Management*, John Oakland (1989, p. 208) states 'processes operated without measurement are process about which very little can be known'. This certainly holds true for workspace changes, where in order to learn about the process of change we need to measure its effectiveness.

Part of the successful implementation of change results from the prior provisions made for future changes. Flexibility in planning and operating premises and their facilities, is of vital importance. However, although flexibility can be considered as a form of insurance in which it is worth investing – through planning for what can be anticipated or even for what cannot be anticipated – it must always be appreciated that it comes at a price. Reviewing the operations of an organization many years after one has been involved in establishing their parameters for new premises, can contribute significantly to the learning process. For example, examine the provisions that were made for flexibility in relocatable partitions on the grounds of the organization's predicted high churn rates. The initial high capital cost would most probably have been justified by projected long-term revenue benefits. However, these benefits are only realized if the relocatable partitions are in fact repositioned prior to the end of their 'natural life'. If, once installed, they are never moved again, then the 'insurance premium' paid can be said to be high. If however, the organization has experienced high levels of churn resulting in many partition moves, then it is likely that the provisions for flexibility will have paid for themselves, possibly many times over. Similarly, audits can assess the use of raised access floors, VAV (variable air volume) air conditioning systems and many others, which will have been justified by reference to provisions for future change. It is of course a fallacy to consider that flexibility can be provided at no cost. Flexibility to permit future change is in fact an expensive commodity. Therefore, where the benefits of flexibility have not been realized the audit should record this for the benefit of similar exercises in the future.

The process of measurement of the effects of the change should be incorporated into the functional audit, which we described in Chapter 3. The post-occupancy evaluation (POE) element of the audit should be considered as a customer satisfaction survey, the aim being to assess reactions to the changes and to identify any remedial or corrective actions that may be required. The process effectiveness element of the audit attempts to establish if the procedures followed were effective and in compliance with relevant corporate and statutory guidelines. Barrett (1995, p. 99) identifies, that there are two types of POE: that is, user-based systems and expert-based systems. In the context of evaluating changes it is our contention that user-based systems are the most revealing, and those which the facilities manager should focus upon, since if constructed properly they can provide a valuable insight into the needs and aspirations of his or her customers. As any marketing professional knows, he who does not consult with his customers is destined to fail. The functional audit has the benefit of 20/20 hindsight, which in the context of examining the effects of change gives us the rare ability to learn directly from the past and, in so doing, look back to the future.

7.1.7 Benefits of good change management

The principal benefits of good change management can be summarized as encompassing the following elements.

- **Identification of responsibilities** – the planning process ensures that all activities are accounted for, the interrelationships between activities calculated and the consequences of certain courses known.

- **Forward planning** – all activities can be planned ahead and appropriate allowances made, contingency built-in, orders placed early and in confidence, and the coordination of all participants simplified.
- **Performance measurement** – the better the project is planned the easier and more accurate the performance systems are likely to be; detailed elemental planning allows detailed elemental performance evaluation; accurate planning = accurate reporting.
- **Data for scenario modelling** – detailed planning assists in the decision making process where the change manager has to choose between scenarios that are aided by forward planning, consideration of 'what if' scenarios as well as assisting with contingency planning.
- **Feedback** – detailed planning and efficient performance monitoring allows the manager to build up a library of knowledge for use on this and future changes, from which the effectiveness of control measures can be assessed.
- **Problem identification and diagnosis** – detailed planning allows the facilities manager to identify problems at an early stage, and can permit trade-off options to be considered.
- **Impossibilities** – planning will give rise to mutually exclusive events or combinations or timings which are impossible, the early identification of which can assist in damage limitation exercises.
- **Sharing the vision** – should encourage a greater degree of information flow within the project organization, which in turn should enable consideration of a wider range of matters, e.g. alternative strategies to be achieved at an earlier stage.
- **Clear management accountability** – a clear schedule of activities provides clear and unambiguous lines and definitions of responsibility, with less chance of misunderstandings and misinterpretations of individual or group responsibility and authority.
- **Improved security of information** – clear organizational and information accountability leads to improved security.
- **Improved team motivation** – most change teams work better and more effectively where the organizational structure is clearly established and where the objectives are clearly defined and programmed.

Case Note 7a

UK government agency. In an assignment, the objective of which was to explore the feasibility of adopting flexible styles of working, detailed consultations were held with the people who worked at the agency to obtain information on their current working practices, activities and locations. The activities of many of the staff involved them being away from their office base for prolonged periods. Until two years previously, all had been provided with a dedicated workplace which, for most, was an enclosed single occupancy office. However, an earlier workspace review identified potential benefits in terms of collaborative working, by having these staff in open-plan rather than enclosed workplaces, which had been met by mixed reactions from senior personnel who considered that 'their office' was an entitlement of their position in the civil service. Consequently, when proposals for the possible introduction of non-dedicated workplaces was presented for discussion with the staff involved, one senior civil servant commented 'first they take away my office, now they want to take away my desk as well'. What is being expressed here is the reaction by someone who sees the only benefits as being corporate and not personal, and for whom the whole exercise will be seen as depriving him of something with nothing being given in return. An effective change management programme needs to address the potential that exists for such issues to emerge as a conscious part of its strategy for implementation and how it is to be communicated.

7.2 Accommodation strategy – maintaining strategic relevance

In Chapter 3 we considered the justification for an accommodation strategy in order to provide a framework for decision-making that guides organizations in the planning and management of their workspace. One of the key responsibilities of facilities managers in managing demand over time is to ensure that the original assumptions that formed the basis of the accommodation strategy are still valid, and remain so, set against the ever-changing business environment.

The provision of workspace is a strategic issue. Used well, space is not purely a cost, but a generator of revenue in its own right. Well-designed and managed space motivates staff as well as attracting and retaining customers. All too often, facilities management is reduced to maintenance, engineering or office services, but space management is a strategic role where the emphasis should be on enhancing the work environment within it, to enable the organization to achieve its goals. In space planning and management, the management of change can be considered at three key levels.

- Changes to the business – its operations, expansion or contraction of the business and its constituent parts, and the identification of business needs as they vary through time.
- Changes to work practices – the adoption of new work methods and processes, the introduction of technology to automate or obviate manual activities, and the changing expectations of the people in the business.
- Changes to workspace – the process of reconfiguring existing workspace, the relocation to new premises, and changes to the way workspace is allocated.

The process of maintaining the strategic relevance of the accommodation strategy is a continuous one. Any review of an accommodation strategy today must consider the implications of the above factors on the existing operational portfolio and the likely future need for other buildings.

7.2.1 Accommodating changes to the business

Measures or initiatives taken to cope with changes to the business in relation to property and facilities, are primarily driven by two main questions.

- Is the space we have, used efficiently?
- Is our overall use of workspace effective?

In general, the former is concerned with unit area per employee in relation to support for the range of tasks performed; while the latter is concerned with output per unit area occupied, and support for the productive capacity of the business. However, it is vitally important for facilities managers to appreciate that the outcomes emanating from these two questions must be strategically driven: 'It is,' as Becker and Steele (1995, pp. 11–12) say, 'about how organizations' leaders choose to convene their employees in space and time in pursuit of long-term competitive edge.'

Against the backdrop of the organization's vision of the desired workplace, feasible supporting strategies will have to be evolved to sustain the relevance of the organization's accommodation strategy, relative to changing business needs. In Chapter 6, the concepts of the strategic facilities brief (SFB) and the service levels brief (SLB) were introduced as essential linking mechanisms between strategic management and operational management. These concepts also provide a basis for the development of an integrated management framework that incorporates the implications of expansion and contraction of the business, upon the operational portfolio and service requirements – Fig. 7.4.

The competitive business environment can no longer sustain, nor afford, getting it wrong first time. This applies to space as much as to human resources, or more importantly, the implications of human resource policies on workspace requirements. The impact of 'downsizing' is a case in point. The pressure of financial performance has had the impact of directing focus on the financial significance of getting the operational infrastructure right first time. An integrated resource management framework that incorporates the facilities dimensions represents a unique opportunity for facilities managers to demonstrate their 'value adding' capabilities by providing viable facilities solutions to emerging business challenges.

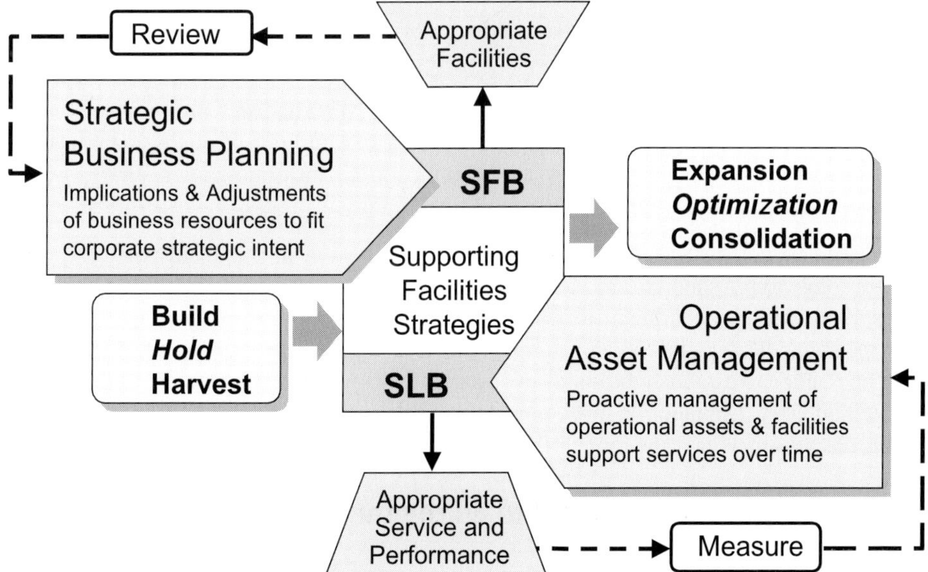

Fig. 7.4 Integrated management framework.

Figure 7.4 illustrates a 'double-loop' integrated management framework emanating from the SFB and the SLB. At the strategic level, the focus of the management interactions between the parties is to review the existing property portfolio in terms of adequacies (e.g. capacity, location, etc) and constraints (e.g. adaptability, component life profile, occupancy costs, etc) in the light of potential changes in the market-place. The outcomes of the SFB being the appropriate facilities mix, corresponding to the current and projected business demands. The 'review loop' is therefore one of maintaining strategic relevance of the operational portfolio within corporate business plans. This strategic evaluation must also take into consideration developments in the supply markets that may impact on the business's overall procurement strategies for premises, support facilities and services.

At the operational level, the focus of management interactions between the parties, is continuously to measure and evaluate the performance of the supporting workspace in terms of the 'value' relationship (i.e. cost and quality dimensions) and 'service' relationship (i.e. quality and time dimensions) to customers (Hronec, 1993, p. 19). The outcomes of the SLB are the appropriate support services and performance levels to meet business needs. As with the review loop, the 'measure' loop covering the delivery of operational facilities services must consider the economics of current provision against potential market offerings.

In general, there are three likely scenarios for an organization – growth, shrinkage or consolidation – each one relating to a different set of factors leading to a corresponding supporting facilities strategy. Typically, the response will be one or a combination of the following actions.

- **Projected demand expected to exceed current supply of workspace** with implications for the need to build new buildings or the leasing of new premises to meet projected demand. This is typified by an organization in an expansion phase, which corresponds to a Build–Expansion facilities strategy, as illustrated in Fig. 7.4.
- **Existing supply expected to exceed projected demand for workspace** with implications for the need to dispose of surplus space or leases. This is typified by an organization in a contraction phase, which corresponds to a Harvest–Consolidation facilities strategy.
- **No significant projected change in demand for workspace** with emphasis being given to sustaining or enhancing the current stock of buildings through modernization or adaptation. This is typified by an organization on an optimization phase, which corresponds to a Hold–Optimization facilities strategy.

It should be noted that reference above to 'supply' relates to the quality of workspace and not solely to its quantity. The issues surrounding, and the decisions involving, the most appropriate workspace strategy to fulfil the corporate intent, are likely to dominate the corporate agendas of many organizations in the near future. Space, as a business resource, must be tracked to establish the status of 'current' supply, guided by an objective demand-led methodology. Demand for space must be established through a clear understanding of the purpose it serves and the range of tasks it is supporting.

7.2.2 Incorporating changes to work practices

Apart from business growth projections, which influence the overall size of the operational portfolio, the other key determinant of demand is the processes of the business.

Organizations are increasingly having to grapple with the pervasive nature and realities of technological development that will directly impact on their work practices. Taken together with fundamental changes in computing technology, communications and rising staff expectations, there will be continual change in how companies approach workspace provision. The resulting demand is about flexibility. It involves building users and how they work, as well as how the workplace can easily adapt over time to changing requirements and the need for different business tools. A central theme of flexibility is that both space and time must be considered and related more effectively to user needs and the functions that are to take place. A consequence of the space–time concept is the gradual demise of the dedicated space for particular individuals. Today, it is not unreasonable to envisage an organization's workplace as a combination of re-engineered space, on-premises options, off-premises locations, and teleworking strategies enabling staff to work anywhere at any time, (see Table 5.1).

In accommodating changes to work practices, apart from the strategies adopted in the provision of alternative work settings and workplace locations, a greater consideration is the human side of change. If the two aspects are not planned as an integrated solution, a smooth transition is unlikely. The workplace ambience is not something that results solely from its manufacture – it is the physical manifestation of the culture of the organization. As such, involvement of the end users in the design of changes is a critical component of the change process, and vital for its success.

7.2.3 Facilitating changes to workspace

Changing an organization is no easy task. Research in organizational behaviour amply supports the view of the organization as a complex, dynamic, living entity whose performance is dependent on deliberate management in order to ensure successful integration and deployment of not only people and technology, but also time and space. The workspace layout therefore represents the physical outcome of the workplace planning process derived from changes to the business and work processes. The primary focus at this level of consideration, is the implementation of the resulting workplace solutions.

From our experience, there are numerous examples of organizations initiating workspace changes where good intentions are not followed through with good execution, resulting in solutions not meeting defined expectations or performance. As mentioned earlier, creating a workplace solution that meets organizational requirements is a multifaceted process that demands a clear methodology to assist decision making, whether the process is one that involves reconfiguring existing workspace or the relocation to new premises.

In Chapter 3 we emphasized the importance of a carefully structured process of data collection, analysis, consultation, communication and feedback to all stakeholders concerned with the design, planning and creation of the workplace environment. Becker and Steele (1995, pp. 155–156) point out several reasons why organizations are not generally good at using effective planning processes in managing workplace transformations. The key reasons for which include the following.

- Poor process-management skills leading to underestimation of the value of teamwork – effective teamwork in workplace planning calls for members to consider each other's values and work styles, but it takes skills and appropriate attitudes that many of us sorely lack.

- The tendency of action-oriented leaders to equate results with tangible 'things', whereas in many situations, success is more dependent on the planning *process*.
- The tendency of organizational leaders to jump to solutions for space problems without adequate diagnosis of the problems and real needs of the organization and its people.

The above blocks to effective planning are symptoms of a lack of clear strategic direction and commitment from senior management to the real issues underlying the promotion of new ways of working. Issues relating to teamwork and collaboration, responsibility and authority, and the perceived role of facilities are all fundamental to clarifying the eventual outcome of workspace design and planning. Without senior management support and a shared vision across the organization, it is unlikely that true transformation of work style and workspace will be successful in terms of matching workplace choices to the long-term strategic intentions of the organization. The absence of a clear strategic direction in workspace initiatives often results in undue attention being placed on the subsequent stages of such projects – the programming, design and construction stages – creating a 'tail wagging the dog' situation. Becker and Steele (1995) refer to the crucial pre-programming stage as the 'Phase Zero' where the focus should be on:

- Testing for agreement on vision, mission and shared values;
- Identifying key themes or directions of change; and
- Defining behavioural objectives – desired ways of working.

In summary, communicating a vision, canvassing users' support, encouraging participation and feedback are all essential ingredients of facilitating successful changes to workspace layout.

7.3 Defining and achieving flexibility

In today's economic and social climate, uncertainty is the norm. Users' tastes, technologies, and organizational structures change at an alarming rate. As a consequence, space cannot possibly be 'future proofed'. The provision of flexibility is essentially an issue of supply and should, as indicated in Chapter 6, be regularly reassessed in the light of future organizational objectives and any new developments in the marketplace. We recommend this reassessment process should be included as part of the annual business planning cycle and annual review of the organization's accommodation strategy. The coverage of this annual review process should be a comprehensive appraisal of needs, taking into consideration the overall space requirements, the space allocations within them, as well as measures to monitor and possibly benchmark workspace to ensure that best practice performance is being achieved.

7.3.1 Defining flexibility and adaptability in buildings

Flexibility and adaptability are terms that are now commonplace in the vocabulary of briefing, building design and space management. With the future being perceived as uncertain, property developers and owners of buildings are increasingly aware, and some are insisting on, flexibility as an essential ingredient. Leaman *et al.* (1998a, p. 1), consider the potential scope of flexibility and adaptability of buildings as covering three aspects.

- The physical adaptability of buildings – constraints arising from the location and site; building shell and fabric, and building services.
- How the property market offers a suitable range of building stock – availability and timing of supply.
- The adaptability of the occupants in 'fitting in' – the facilitation of change in work practices and processes, in workspace layout, and the provision of workplace choices in promoting the organization's culture.

In a wide-ranging discussion of the topic, the authors distinguish the characteriztics of flexibility in buildings as 'more about short-term, often potentially reversible, changes of relatively low magnitude; adaptability with in-built potential for large-scale changes of greater magnitude in the longer term. ... Adaptability involves additional knowledge of context, purpose and application ...' Leaman *et al.* (1998a, p. 1)

In space management just as it is in business management, flexibility embraces the evolving relationship between the individual as an employee, and the business as the employing entity, and their interface within the work environment. The need for business agility in a dynamic global environment, the increasing trend towards a 'project focused' structure of working and development of flexible workspaces, all contribute towards a continuous evolution of the individual–business interface. The continually evolving dynamics of work and work processes as drivers of demand within businesses are a subject we discuss in Chapter 8. Table 7.1 summarizes the various components of flexibility in such a context.

7.3.2 Achieving flexibility in supply

Flexibility in supply is a variable of the property market's offerings in terms of the timing of availability, range of building stock and quality of workspaces they contain. In the context of an organization, we are concerned primarily with the supply of appropriate workspaces to accommodate a range of tasks that businesses need to carry out in order to produce their respective goods and/or services.

As relationships between business performance and space use become more obvious, the practice of space planning will become more sophisticated. The emergence of space planning went hand-in-glove with the growth of facilities management. As office activities intensified and diversified, so did the spaces and services required to support them. As a result, facilities managers have to cater for an increasingly technologically driven *push* to provide alternative facilities solutions to meet emerging business needs.

In Chapter 5 we considered the development of alternative officing strategies as a response to optimizing the space–time equation. The factors driving the rapid pace of changes in workspace use include a combination of shifts in mindset in organizational strategic thinking and structuring, competitive pressures, uptake of technological tools and systems, and changing lifestyle and work processes. Leaman *et al.* (1998b, pp. 1–2), provide a good summary of the social, technical and economic infrastructure which produces the preconditions for change.

Table 7.1 What flexibility means to the business and the individual

Component of flexibility	Business	Individual
Workforce	Balancing labour with the peaks and troughs of production in response to varying customer demands for goods and services	No long-term commitment to work for a single employer, requiring alternative sources of work to be actively canvassed
Workspace	Minimization of fixed costs of real estate and workplace infrastructure commensurate with the provision of facilities that enable the workforce (individually and collectively) to be effective	Working from a variety of locations to suit the prevailing needs of businesses (including home wherever possible) with no dedicated workplace within the 'employers' premises
Work hours	Whatever timing and duration is necessary to meet customers' needs	Variable by day, week, month – to suit the combined demands of different 'employers' and their work patterns
Skills	Accessing skilled specialists only when required, while routinely operating with a mutli-skilled workforce	Continuous development of personal skills, both specialisms and range
Property	A combination of core and peripheral accommodation operated under flexible short-term agreements	The provision of a workspace in the home supported by appropriate infrastructure, with access to a shared neighbourhood telecentre
Environment	Readily and cost effectively reconfigurable	Conducive to productive working and readily adaptable to personal needs
Facilities management	Effective management of all the means of conducting business and the interfaces between each	The provision of a composite range of support services to sustain all work needs

- More heightened awareness of business activities being affected by the layout and performance of the working environment.
- Greater strategic awareness of the importance of buildings as a resource for the organization as a whole, perhaps stressing their benefits more than their costs.
- Growing dependence – now almost total – on the networking, distribution and storage capabilities of information technology.
- More developed emphasis on getting the best out of people at work, especially if it is physically and mentally demanding, critical to the organization or unduly stressful, risky or dangerous.
- Changes in working conditions, with statutory drives for fewer hours at work, more flexible working, variations to work contracts and instability in labour market.
- Uncertainty in business futures, especially in traditional businesses like banking which are challenged by changes induced by information technology.
- More emphasis on knowledge work, as opposed to other types of administrative or clerical activity.
- Pressures towards more group and project work, with growing importance of the workgroup.
- More personal freedom and autonomy.

- Less time spent at desks, and more in meetings and other activities away from the desk.
- Greater mix of activities.
- More intensive use of space, but also greater diversification, with the relentless trend towards more remote working.

In space planning terms, the above developments or pressures have spawned a number of new perspectives of space use classifications based on new dimensions like: the nature of work and the nature of change (Duffy *et al.*, 1993); occupancy and interaction (Leaman *et al.*, 1998b); interaction and autonomy (Laing *et al.*, 1998). These and other models have given rise to a variety of physical layouts or workspace settings. They have also led to new terminology used in the layout of workplaces with names like Teamspace, Office 2000, Right Space, plus exotic names like huddle rooms and bullpens.

More recently in the UK, a new dimension of supply flexibility has emerged in the form of fully serviced offices with inclusive and highly flexible terms. Currently, a minor segment of the property market, modern serviced office space, is forecast to be a major growth sector (Jones, 1997). The concept of the serviced office or business centre, originated in North America during the early 1960s, but it was not until the late 1980s that the benefits it offered were recognized in Europe. As worldwide recession overwhelmed economies everywhere, companies of all sizes were forced to look aggressively at cutting costs. Serviced offices were identified as an attractive alternative to owning property assets, enabling companies to convert fixed costs into variable ones and release capital to use within the business. In addition to cost benefits, the serviced office option also offers companies the flexibility and mobility of arriving at a location whenever they wish and remaining there however long they need to stay, backed by uncomplicated and easy-to use services.

7.4 Property procurement and service agreements

The nature of building as a capital product dictates a financial commitment as a long-term investment. The long-life durable nature of buildings as a group of assets calls for a continual period of management, maintenance and reinvestment in order to sustain their value and utility. Its locational rigidity and site constraints make each building unique, offering a diversity of supply and quality that caters for the needs of different types of businesses. The time lag between the decision to build and occupancy results in a procurement–delivery cycle that is often protracted and susceptible to the vagaries of the market. At the same time the future use of buildings cannot be forecast with certainty.

As an essential operational asset supporting and accommodating the business's productive processes, the issue of ownership options is clearly a matter of significant financial importance, given that workspace can be an organization's second most expensive resource. In recent years, the trend for traditional preferences of owner-occupation and long-term leasing has given way to alternative methods of procurement, both in the building as well as the support services that make it a fully functional workplace. In this section, we consider the issues of procurement of operational buildings and their associated facilities support services.

7.4.1 Changing view of property ownership

In the procurement and management of operational property, there has been a distinct shift in focus from 'supply management' to 'demand management' (Chekijian, 1996). Since the 1970s, the role of property management within corporations has been recognized as one of defining capabilities and performance measures from the supply side of the property industry. It rode the tides of the property booms and slumps over the two decades that followed, latterly with a focus primarily on value rather than on cost. Supply management emphasis involves the responsibilities for managing the supply of operational property, where the focus has been on management of the facilities provision, the traditional arena of the property or real estate management function.

Since the late 1980s, demand management has prevailed where the focus is to manage affordably the consumption (utilization) of the corporate property portfolio. This transformation in process is a significant shift, in that it signals a change in the competence profile necessary for facilities managers of the future, a subject we will consider in Chapter 8. This has been substantiated by the findings of the Corporate Real Estate 2000 research programme, which has shown that facilities managers have to begin to shift their focus from cost reduction and containment to productivity improvement and growth. A number of financial-economic drivers have been responsible for this shift from a supply-led to a demand-focused process.

- **View of ownership of property assets** – Competitive pressures have forced many companies to refocus on what they do best by shrinking back to their core business(es). Business re-engineering initiatives have led to a spate of downsizing and outsourcing of non-core support functions. For many companies, like IBM, the re-engineering exercise has resulted in a legacy of surplus or inappropriate property assets, such that a strategy aimed at releasing capital tied-up in property assets and the associated reduction in occupancy costs, became an important element of the overall business recovery process. The increasing availability of serviced workspace offers an attractive option to traditional ownership or leasing with its management and administrative overheads and FRI (full repairing and insuring) constraints.

- **The economics of contracting out** – the trend towards outsourcing of property and facilities services has gained considerable pace in recent years. The conceptual basis from which the practice of outsourcing has been derived, has its roots in the writings of Peter Drucker, Tom Peters and those listed in Table 2.2. In the main, two principal strategic approaches formed the basis for procurement evaluations by companies in the sourcing of support services: first, concentration of the firm's own resources on a set of 'core competencies' where it can achieve definable pre-eminence and provide unique value for customers; second, strategically to outsource other activities – including many traditionally considered to be integral to any company – for which the firm has neither a critical strategic need nor special capabilities (Quinn and Hilmer, 1994). The implicit assumption is that when intelligently combined, core competency and extensive outsourcing strategies can provide improved returns on capital, reduce risk, improve flexibility, and provide better responsiveness to customer needs at lower costs. The emphasis on the sourcing decision of non-core functions as a strategic evaluation process, once again highlights the need for senior management guidelines as to what are acceptable risks associated with the placement of any hitherto in-house function, to an external service provider.

- **Drives to improve occupancy cost management** – An area of strategic evaluation that has raised considerable attention from senior management in recent years is how real estate decisions impact on company profitability. The focus of financial analysis with the intention of demonstrating that real estate decisions can have a large impact on a company's profitability, has largely centred on scrutiny of occupancy costs. Because it is relatively recently that business managers have become aware that costs associated with operational real estate can typically account for 10–15 percent of the total operational costs of a company, profiles of occupancy costs have only started to be assembled in recent years. Work done by Apgar (1993, 1995), Evans (1994) and Chekijian (1994) demonstrate the power of systematic analysis and the need to 'test' alternative scenarios in the strategic evaluation of property and facilities support costs. They also show that corporate efforts to control costs and improve profitability lie in the efficient use of space.
- **Trends to internal recharging of property and facilities services** – For facilities managers, the practice of internal charging, for the first time, places the onus on them to define and cost the services they provide to their 'internal customers' (i.e. the users of the services within the organization). More significantly, the shift forces in-house facilities departments as service providers, to move from a culture of reacting to demands from business units as their *raison d'être*, to one in which the service culture governs the transactions between service provider and the customer. In such a situation, the ownership of space costs is placed squarely on the shoulders of business units as the purchasers and consumers of the services.

7.4.2 Development in service level agreements

The overall process of the provision of corporate operational facilities must be carried through to the efficient provision of facilities support services if the benefits from real estate resources are to be fully realized. This will involve the evaluation of the appropriateness of the current range of facilities support services in terms of costs associated with their current provision and the level of service provided. The objective is to ensure the best fit of service packages in support of the core business. In this respect, the assessment of demand for support services is an essential element of a well organized and relevant facilities and services provision. Such an arrangement can provide valuable management information for understanding the role of support services and the issues relating to their efficient management and timely supply.

The concept of defined service levels in property and facilities provision is relatively new. The promotion of their use is a direct result of the implementation by in-house service departments of internal re-charging to business units, in both private and public sector organizations. This much-needed change in culture has been built upon negotiations between the parties concerned about the required level of service provision, which is directly related to the cost and quality of provision. There is evidence to suggest that where organizations recharge the costs of their workspace and facilities services to the work groups that use them, this is likely to engender prudent usage. This is, however, not always the situation. Internal recharging of workspace is really only likely to reap the desired results where the work groups concerned have financial criteria included as part of the evaluation of their overall operating performance. Where no such financial criteria are routinely applied, then recharging for workspace and facilities can become an academic exercise, since the significance of their cost is unlikely to register with the work group managers.

Hiles (1993) uses a pyramid model to define service level and value. The model identifies four levels of service from Level 0, where the customers are on their own, to Level 3, the highest standard of service – Fig. 7.5.

Level Zero may apply to areas where customer expertise exceeds that of the service supplier, or to non-critical areas. Level 3 may apply to mission-critical or high-value services. In general, the higher the level of service, the higher the costs and the lower the risk of loss of service: the lower the level, the cheaper the service but the greater the risk of loss of service. In evaluating individual service demand, management is prompted to establish the quality of service appropriate to its needs, as opposed to its wants. In other words, the evaluation process defines the minimum acceptable service requirements. Using the above model, Hiles (1993, p. 2), defines a service level agreement (SLA) as: 'an agreement between the provider of a service and its customers which quantifies the minimum quality of service which meets the business need.' Some key words within this definition require qualification.

- It is *an agreement* – it is negotiated and involves a growing understanding of the needs and constraints on each party (customer and supplier), probably resulting in compromise.
- It *quantifies* the level of service – metrics are designed which both parties to the SLA agree represent the quality of service to be delivered and its commensurate cost.
- Delivered quality is the *minimum acceptable* – anything above the minimum may be excessive and therefore probably result in unnecessary cost, but the service delivered has to be acceptable to the customer.

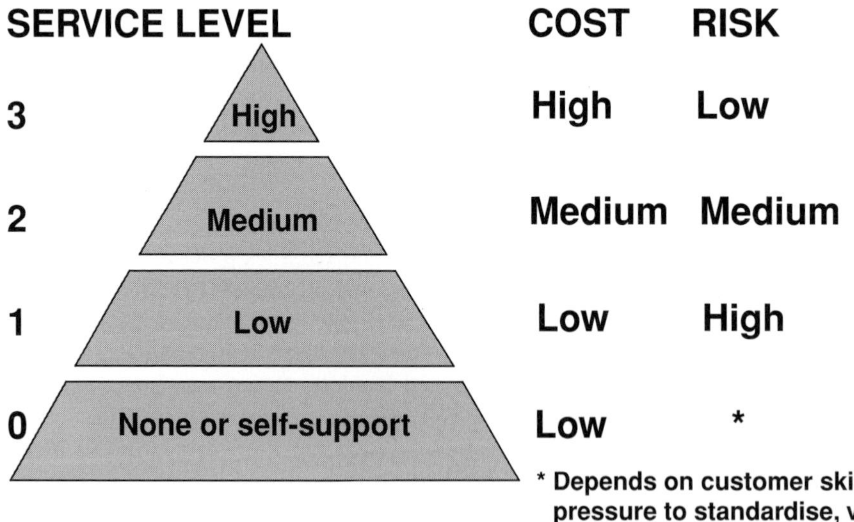

Fig. 7.5 Service support characteriztics (Hiles, 1993).

Hiles' definition of SLA explicitly states that the appropriate service level that management will opt for is the minimum level. Pratt (1994, p. 10), provides a wider definition of service level agreement by suggesting that there is an element of choice that provides for both increases and decreases in the level of provision, stressing that flexibility is essential if the business is to be able to respond quickly. 'A Service Level Agreement is,' says Pratt, 'a statement of various service level options from which one will be selected by the customer or client which specifies timing, frequency, cost, etc to match the business need.' Both of these definitions concur on the requirement to meet or match service provision to the business need. For many organizations, their services have been adapted and enhanced over many years, during which business objectives, market conditions, priorities and cost-benefit calculations have changed radically. But even when business goals are shifting, an SLA provides the impetus to review such services against the prevailing need, particularly to isolate 'custom and practice' and identify the *real* current service requirement.

7.5 Churn and controlling it

We are in a period of change driven by technological innovations and organizational restructuring, which is already challenging the conventional concept of the static individually-assigned workspace. The convergence of technologies in computing, land-based and satellite communications, data and visual transmissions, is offering new possibilities of where and when we choose to perform our work tasks. Seemingly, for many, choice and mobility are parameters that are added to place and time. As a result, flexibility in provision and use are becoming essential elements in space planning and space management. For the facilities manager, this further flexibility comes at a price – the added costs of providing for enhanced flexibility and mobility. A consequence of having to provide for and to accommodate this added flexibility at the workplace, is the need to define and assess the nature, and the rate of potential movement of users within and between buildings.

7.5.1 Definitions and nature of churn

For any organization, as staff or teams of staff are moved, cut, merged, added to or transformed, space changes and adjustments are made on an almost continuous basis. *Churn* is a term given to the relocation of staff of an organization within a building or between buildings. Churn can be a consequence of an organization's culture and work practices. An obvious example is the promotion of a team-based culture, which necessitates the constant formation and re-formation of teams of people from different parts of an organization for varying durations of time. Movement of people as a result of organizational restructuring or re-engineering initiatives are other examples that add to the amount of churn within a business. *Churn rate* is defined as the total number of employee workplace moves made in a year expressed as a percentage of the total number of workers in the premises. Hence, a churn rate of 100% implies that, on average, each employee within that organization has moved to a new location within the same building, or to another building, within the span of one year.

Benchmarking data produced by The Workplace Best Practice Group, indicates that churn rates in the UK range from 20% to 90% per annum, with the majority of businesses falling between 30% to 50% – with a sizeable attendant expenditure when the cost per workplace move can be of the order of £2000 per person moved.

Incorporating or providing flexibility in workspace design is a common strategy used to accommodate churn. However, there are various dimensions to the concept of flexibility when applied to the design of workspace layouts and configurations. Thatcher and Thatcher (1996, p. 2), in their guide, *Reducing the Cost of Churn*, defined flexibility as 'the capacity to adjust and modify, in some way, to respond to a change in activities or functions'. They also described the four dimensions of possible adjustments shown in Fig. 7.5.

- **Versatility** – the ability of the space to handle different functions without physical change – universal footprint planning, where the only change would be the files brought in by the people doing the work.
- **Rearrangeability** – the ability to cope with change by moving and rearranging equipment or furnishings, such as storage, worksurfaces, tables, lighting, within a given room or space, as it is the way they are laid out that provides the flexibility to support different activities.
- **Convertibility** – this method of managing change is to reassemble interchangeable sub-components into new configurations or new functional assemblies, which use system furniture, modular walls and modular utility systems.

The main differences between convertibility and rearrangeability are scale and the number of building elements involved.

7.5.2 Managing churn

A feature of modern organizations today is the ability to respond to customer requests and market challenges. In addition, fierce competition is dictating that the ability to respond is insufficient in itself, as the speed of response is also becoming critical. In this respect, while change is the watchword, so churn must follow. A major responsibility for the facilities manager is to manage the types of churn as efficiently and effectively as possible.

Every workplace move bears a cost in terms of the direct costs of the moves as well as the indirect costs of associated disruption. In this respect, it is necessary to categorize the types of moves in terms of their cost implications and planning considerations. A *briefcase move* is defined as moving people from one existing workplace to another existing workplace. No furniture is moved, no new wiring or telecommunications are required, just support documentation and supplies are moved. A *furniture move* includes reconfiguration of existing furniture and/or furniture moved or purchased. No new wiring or telecommunications are required. A *construction move* includes new walls, new or additional wiring or telecommunication systems, or other construction. Clearly, each of the three types of moves has very different attendant costs. Briefcase moves being the most economical, followed by furniture moves and construction moves respectively. In assessing the potential cost of churn for the organization, the facilities manager must assess the likely profile of types of moves against the size, mix of groups and their work processes. Measures to control and accommodate churn must be planned and, whenever opportunities allow, integrated with the organization's accommodation strategy. In general, initiatives aimed at reducing the cost of churn fall into two options:

- Reduce the rate or frequency of churn, and
- Reduce the cost per move.

Reducing the churn rate will reduce the cost of churn. However, churn is a result of an organization's need to respond to the changes in its markets and may be a product of its work processes and culture. In this respect, it may be counterproductive or impractical to reduce the cost of churn by reducing churn itself. However, it is the role of the facilities manager to monitor the costs and indicate to business managers the impact of the churn on facilities operating costs, and hence the cost of being in business.

For the facilities manager, measures aimed at reducing the cost per move offer more avenues for action. However, senior management's endorsement is a precondition, as any initiative will impact on the design outcome of the workspace layout and allocation. An underlying challenge is to develop solutions to avoid the expense of planning customized layouts and design each time a work team changes its organization and/or size. Measures aimed at reducing churn cost include the following.

- The adoption of a set of universal footprints or templates – so that any churn is confined to only briefcase moves.
- The provision of a grid throughout the workspace, in which workplaces of different types can be located and changed with the minimum of disruption to and reconfiguration of, services.
- Having fewer space standards, with fewer differences between each type, in terms of their fixed components.
- Where functional differences are required to be addressed by workplaces of different configuration, these differences are achieved by modifications to a generic 'core' workplace rather than the replacement of one complete workplace setting with another – this is likely to be achieved by the use of component-based furniture systems, rather than the use of a monolithic furniture product.
- The servicing of workplace settings through the use of standard IT desktop configurations, where differences of user specification are achieved through networked software applications instead of different hardware.
- The adoption of cordless networks in place of hardwired networks, with the corresponding reduction in both cabling and the complexity of moving workplaces.
- The use of technology systems to enable effective routine communication to be achieved, such that the driver for co-location of people is based predominantly on their need for face-to-face communication and interaction.
- The use of local and central document storage systems, which possess the capability of being relocated safely while fully-laden.

7.6 Improving utilization

The management of workspace requirements over time, starts with the assessment of needs. So how much space does your organization need? How much? Ah yes, but how much space does your organization *really* need? That is a tough question to answer. Many organizations may know how much space they require to meet the needs of the business today, but how about next year? And the year after that, and then the one after that, and then the one after that?

The reason it is a difficult question to answer is that it is bound-up with the forward planning of the business which, as we all know, is an extremely difficult process. So what organizations are really saying is that because they find it difficult, if not impossible, to forecast the business well in advance, then they can hardly be expected to forecast their accommodation needs well in advance. While this *may* be true, it does not stop businesses from making major commitments to property, freehold and leasehold which are decidedly certain, i.e. they own the property or are committed to a 25 year lease. What is needed, therefore, is a means of addressing the flexible requirements of the business, with a facility (i.e. property) and the way it is used, which has traditionally been inflexible. The starting point is to understand what the business has today by way of property and workplaces and how well they currently serve its needs; after which, the facilities manager can then move on to assess if and how they may best serve business' needs in the future.

Effective utilization of the premises is determined by the suitability of the environment for the work tasks to be performed. In addition however, it will also take account of how personnel are deployed within the building. Work practices and layouts that are inefficient in their use of space clearly lead to an underutilization of the facilities. However, it is also important that space liberated as a result of efficient planning is grouped such that it can be released for other purposes, either within the organization or by other parties. In instances where breaks have been engineered into the property lease, then the gathering together of this surplus space could permit disposal of non-performing assets. However, the success of such exercises is dependent upon the ability to permit the liberated space to be used by another occupier, either through disposal, assignment or sub-letting. For example, are the services configured such that consumption can be separately metered? Such requirements, together with access, security and use of central services, can all impact upon the viability of disposal plans.

The premises audit referred to in Chapter 6 will provide a measure of how efficiently space is used. Workspace utilization is not only a function of the skill of the space planner in creating an efficient layout, but has been largely derived from the degree of flexibility the premises afford its users in terms of the scope for repositioning workplaces. The utilization factor also possesses a time dimension, i.e. for what durations are the building(s), and the workplaces contained therein, occupied? There are 168 hours in every week and that is the time the real estate owned or leased by the organization is available. Yet for most organizations it is probably only open for use for 50 hours a week, and then only occupied for 30 hours a week at best. To put it another way – the building is at best utilized for 18% of the time and that is assuming that staff are always at their workplace. Other methods of using workplaces on a non-dedicated or time-shared basis could have some benefit, an aspect that is referred to in Chapter 6.

7.6.1 Multi-sites issues

The issues of applying controls and operating effective premises procedures are compounded where an organization requires to implement consistent standards across a range of operational sites. Consistent application of standards is to be encouraged, because of the benefits of cost effectiveness and easier management. Consistent standards, both in the allocation of space and the provision of facilities, are aided (and in some instances only made possible) by developing a corporate facilities handbook in which guidelines for procedures are defined. The guidelines would include procedures to be followed for the budgeting and the allocation of space as well as for implementing and controlling future changes. They would describe and illustrate each corporate space standard, detailing the furniture, equipment and services required for each function. However, the handbook and its contents are only of value if they are kept up to date. Regular and effective feedback from each site management team is necessary to ensure that premises and their facilities satisfy the requirements of statutory and legislative compliance, meet current demands and benefit from experience in application. The fundamental operating principle for the premises themselves is the same as for developing and using the handbook, namely – change is the steady state, and all aspects of workspace must be reviewed in this context.

7.6.2 Developing workplace solutions

First, it is important to recognize that there is no single correct solution. At best there may be several solutions that might be appropriate for an organization to implement, and these are likely to change through time since they will need to 'flex' to suit the changing circumstances of the business. If the facilities manager is particularly unfortunate it is possible that he or she may be faced with only one solution. In our experience, there is never a situation that has no solution. There are likely to be those that it may be difficult to implement, or require tough decisions to be taken, but a solution always exists. The issue here is to what extent the business could or should apply the approach of non-dedicated workplaces, a subject considered in Chapter 6. In the context of improving utilization, non-dedicated workplaces can have an important part to play, although such flexible working arrangements are intended principally for improving the performance of people. Where they are applied solely, or even principally, as a means of achieving improved utilization of workspace then extreme care should be taken. The primary concern with all such approaches as address-free, JIT (just in time) and the virtual office, is that they depend for their success on a departure from the dedicated allocation of space. No matter how cost effective and superficially beneficial this may appear, there is a principal challenge to its acceptability. The human animal is territorial and environmental research over the last 50 years has indicated man's need for a hierarchy of defensible space. Given the short-term nature of the application of these new concepts, it would be foolhardy to discount the evidence of several decades of research. Consequently, non-dedicated workplace provision should only be adopted as a result of extensive investigation into its suitability, and then only where it will improve the productivity of the people concerned. Notwithstanding that 'health warning', where it is founded on a thorough understanding of the needs of the people in the business, and is supported by appropriate workplace provisions and services, then flexible styles of working can be very effective.

7.6.3 Workspace disposal

If after exploring all feasible options, it is decided that the premises are no longer of any use to the business, then the question should be asked: why should the property be of any use to other businesses? Is the reason the building is redundant because:

- It is in the wrong location?
- It is no longer serves business and work process needs?
- It is the wrong size – too big or too small?
- It is too inflexible?
- It is no longer economic to operate?

If the answer to any of these is yes, then the facilities manager has to consider why other businesses should consider the property to be an attractive proposition? And what can be done to address these failings and make the building suitable for other potential occupiers? The solution may lie in addressing some of these issues to make the premises an attractive proposition for others, and in some instances may even make the building worth a 'second look' by its current occupiers!

In developing possible solutions, all parties involved need to chant the mantra: 'there are no sacred cows'. Sometimes the best attribute of a building, i.e. the reason the business wants to retain it, is the very reason someone else will buy or lease it, so the facilities manager has to be prepared to think and act flexibly. Not to the point of 'self-destruction' but if oversupply or ineffective utilization really is hurting the business, then 'what price a pain reliever?' Even when it appears that you are 'locked into' a property agreement, there are always possibilities that could and should be explored. Although the detail will be specific to each organization and its property, the following are intended as a generic guide to some aspects that could be worth considering.

- **What is the real value of the property?** – The value of the property to the current occupying business may be entirely different (more or less) to its value for others. Facilities managers should beware of situations developing where the business will hold out indefinitely for an unrealistic price for its property, in the mistaken belief that there is someone 'out there' who would be willing to meet the value at which the business has the property in its books. Conversely there may be someone 'out there' who, through their use of the premises, could realize more value in the property than the current occupier could do.

- **Releasing the value of the property** – Although the business may be seeking to off-load the property because it is a cost burden, the answer to realizing and releasing the potential value of the property may lie in making an investment in its development or enhancement, thereby unlocking its innate value. The facilities manager should avoid the trap of thinking that such initiatives are 'only putting more good money after bad'.

- **Is a partnership possible?** – Rather than selling or otherwise disposing of its whole interest in the surplus property, could the business enter into a partnership with another party, in which both share in the upside of any initiatives – such as the operation of shared facilities or the marketing of serviced workspace?

- **Why go it alone?** – Instead of thinking that the business has to find the solution itself, is a joint venture with another party a possibility to releasing the asset value, through development of the premises (refurbishment, conversion or extension) or the redevelopment of the building's site?

- **Are the premises a revenue cost or an investment?** – Just because the premises are no longer suitable for the work needs of the current occupier, should not rule out their potential use for other businesses for whom the premises could be a significant improvement upon their existing work environments. Consequently, consideration should be given to the potential that exists in disposing of the premises as *operational* assets, while retaining them as *investment* assets.

Where a prolonged period of non-occupation is envisaged then 'de-rating' of the property may be appropriate. This involves stripping-out elements so as to render the building unusable, and hence not liable to local rates or taxes. However, while this may seem attractive in terms of an immediate reduction in outgoings, remember that to make the premises suitable for another occupier to use, then these facilities will have to be reinstated, which will undoubtedly cost more than to remove them.

It may be possible to redevelop the property for other uses or mixed use (Fig. 7.6). Where this is being considered, an early consultation with the local Planning Authority is advised, in order to identify what plans are proposed, what zoning classification has been allocated to the area and to ascertain what may change in the future.

As an interim or long-term measure, the operation of shared space and facilities could prove beneficial, although not to the extent that it distracts from the core business – although the organization is already likely to be distracted by the cost of this 'white elephant'. If sharing premises with other businesses is being contemplated, then one needs to give consideration to the following.

- **Building Services** – in order to permit the building or parts of it to be sold or let independently, environmental services will have to be isolated with independent supplies and metering of workspace by zone, floor or wing of the building.
- **Access** – entrance(s) for personnel and goods.
- **Lifts and stairs** – providing access to all floors or segregated.
- **Security** – central or by building zone.
- **Car parking** – allocation and 'policing' of spaces.
- **Common services** – site maintenance and utilities.
- **Charges** – the basis of charging for shared facilities and services needs to be determined – by actual consumption is favoured, in preference to the area occupied or a per capita basis.

Of course these principles do not just apply to empty space, they can also be applied to underutilized elements of accommodation. For example, on a business park where an organization had reduced its staff numbers significantly, the central services in their building were correspondingly over-sized for the remaining staff. By applying the principles above, and with some innovation, they now open their canteen facilities to staff of other organizations on the business park and hence mitigate their costs.

Finally, facilities managers should approach the whole exercise of property surpluses and disposal with an open mind and challenge accepted principles. Remember, whatever

(a)

(b)

Fig. 7.6 (a) and (b) The conversion of buildings from one use to another can make good economic sense as well as good environmental sense. Bonded warehouses converted to offices at Commercial Quay, Leith, Edinburgh.

the issue, it can be negotiated out, and for the right deal the Landlord may well be prepared to take back the property. What seems to be a high price today, may appear insignificant in the long-term when compared with the impact the retention of inappropriate premises and under-performing assets can have upon the business.

References

Apgar IV, M. (1993) Uncovering your hidden occupancy costs. *Harvard Business Review*. May–June, 124–136.

Apgar IV, M. (1995) Managing real estate to build value. *Harvard Business Review*. Nov–Dec, 163–164.

Barrett, P. (ed) (1995) *Facilities Management – Towards Best Practice* (Oxford: Blackwell Science), p. 177.

Becker, F. and Steele, F. (1995) *Workplace by Design – Mapping the High-performance Workscape* (San Francisco: Jossey-Bass), pp. 11–12.

Chekijian, C. (1994) Shrink, shrink, shrink that space for big savings, increased profitability, shareholder. *Corporate Real Estate Decisions*, 7 October.

Chekijian, C. (1996) Demand management must focus on corporate objectives. *Corporate Real Estate Executive*, **11** (2), 36–37.

Duffy, F., Laing, A. and Crisp, V. (1993) *The Responsible Workplace – the Redesign of Work and Offices* (London: Butterworh Architecture in association with Estate Gazette).

Evans, M. (1994) The impact of real estate decisions on profitability. *Site Selection Europe*, March, 28–30.

Hiles, A. (1993) *Service Level Agreements – Managing Cost and Quality in Service Relationships* (London: Chapman and Hall).

Hronec. S.M. (1993) *Vital Signs* (New York: AMACOM) p. 19.

Jones, O.J. (1997) The service office – the future of flexible space? In: *Proceedings of the British Institute of Facilities Management – Facilities Management Into the Business Revolution*, Cambridge, 15–17 September, pp. 31–38.

Laing, A., Duffy, F., Jaunzens, D. and Willis, S. (1998) *New Environment for Working – The Re-design of Offices and Environmental Systems for New Ways of Working* (London: E & FN Spon) p. 15.

Leaman, A., Bordass, B. and Cassels, S. (1998a) *Flexibility and Adaptability in Buildings: the 'Killer' Variables* (London: http://www.usablebuildings.co.uk).

Leaman, A., Cassels, S. and Bordass, B (1998b) *The New Workplace: Friend or Foe?* (London. http://www.usablebuildings.co.uk).

Oakland, J.S. (1989) *Total Quality Management* (Oxford: Heinemann Professional Publishing).

Pratt, K. (1994) Introducing a service level culture. *Facilities*, **12** (2), 10.

Quinn, J.B. and Hilmer, F.G. (1994) Strategic outsourcing. *Sloan Management Review*. Summer, p. 43.

Thatcher, C. and Thatcher, G. (1996) *Reducing the Cost of Churn*. Facilities Management: Management Guide No: 10 (London: The Eclipse Group).

8

The future of workspace management

The provision of workspace, and hence its management, should be a direct response to the needs of people (individually and collectively) in supporting them in their work endeavours. It follows therefore that as the requirements of work itself change, then so too will the requirements of the management of workspace in response to people's different work needs.

In the preceding chapters we have concentrated our considerations upon methodologies that can be applied to the management of a business's workspace – that is, the reconciliation of the *demand* expressed as an assessment of work needs and the *supply* in terms of the building stock that is available to the business. Of course, this has involved assessing the needs of today and tomorrow, and in so doing considering the future in terms of both the needs of the business and the provisions of the property market. But what is the future of the management of workspace itself? Is there a future need to manage workspace, or at least will the management processes of the future be the same or different from those of the present?

In this chapter we turn our attention to considering the future role workspace management will, or should, play within business operations. Inevitably the future of workspace management will be influenced, if not determined, by what the future holds for the component parts of the space management process, namely: work and work processes; the people who comprise the workforce, their lifestyles and expectations; the technology of work, affecting both the users of the workspace as well as the process of its management; and the property industry and its responses to the dynamic needs of those it serves. We will therefore explore what the future holds for: work, the workforce, lifestyles, technology and property; and postulate, if these changes occur, what impact they could have individually and collectively upon the role of facilities managers and the process of managing a business's workspace.

8.1 The future role of work

Workspace – its location, its attributes and its effect upon us – is dependent upon the very existence of work, its nature and its needs. It is important therefore to consider the likely future role work will play within society at large; the way society, business and individuals reconcile their need for work; how and where that work is to be conducted; and the impact these decisions will have for the lifestyles of us all. In giving consideration to such issues, we draw heavily upon the seminal work in this area by the RSA in *Redefining Work: An RSA Initiative*, by Valerie Bayliss (1998).

'Change is inevitable. In a progressive country change is constant.' So said Bejamin Disraeli, the former British Prime Minister, in a speech he gave in Edinburgh on 29 October 1867. His words could, however, so readily apply to the 'trans-millennium' age we currently inhabit. The principal difference is that today the nature, extent and speed of change are greater than ever before – which most probably could also have been said at any point in history – since the index by which such events are measured is the past experiences of the individuals concerned. So what changes does the future have in store for work? The future, in this context, is 20 to 25 years hence, that is in the range 2020 to 2025. Then, the world of work will be much, much different from the world we know today. What has been regarded as the accepted model for working life – the 40/40 model, 40 hours a week for 40 years – is in fact a comparatively recent phenomenon, stemming from the era of mass production and centralized bureaucracies of the 1930s. While this still bears some resemblance to the way many people work today, it will have no place in the future world of work. Bayliss (1998) describes the vision the RSA have of this world of work in 2020 as comprising 12 elements, from which we draw our picture of the future world of work.

- **Business convergence** – where the distinctions between public and private sectors, between one business and another, between banks, supermarkets and schools are blurred if not obscured or irrelevant.

- **Careers** – are no longer thought of as single, continuous, life-long pursuits but are fragmented and intermittent with individuals moving between several 'employing' organizations (public and private), working in different occupations throughout the span of their working lives.

- **Structure of organizations** – moving increasingly towards the virtual organization, where many private, public or voluntary organizations will no longer produce goods or services themselves but instead orchestrate the actions of others, many of whom will never actually see the end-product.

- **Jobs** – the concept of a job will change from being a bundle of fixed tasks and responsibilities to the development of roles where people (singly and in concert with others) apply their skills in a variety of ways to suit the changing demands of the organization; work will increasingly focus on the exploitation of knowledge in many diverse forms.

- **Employment** – the distinction between employment and self-employment will become increasingly irrelevant as people operate in a variety of ways, 'selling' their skills via many different arrangements with businesses, often several different ones at once, as they work 'for' themselves rather than 'for' someone else.

- **Working hours** – fixed working hours set by an employer will be replaced by people working when and for as long as they need, in order to achieve a predetermined output.

- **Place of work** – the role of the 'employer' will be the person who has secured the business for a given output and not someone who demands attendance at a fixed time and place; most workers will split work time between a range of locations which are likely to encompass a central base, a service base and home.

- **Pattern of work** – 'continuous work', although still existing for a few, will give way to what has previously been thought of as self-employment, where people move from 'project' to 'project' rather than working for one organization on one activity for a prolonged period.
- **Technology** – far reaching developments in information and communications technologies will penetrate sectors and jobs that were untouched during the 1990s, so much so that almost every job will be changed accordingly, and skilled 'e' workers will be the accepted norm.
- **Business travel** – encouraged by the environmental unacceptability of pollution emanating from transportation, business travel for many will be reduced if not obviated by the use of technology and systems that will enable people to work together simultaneously, independent of geography – local, national and international – with business travel only being required where face-to-face interaction is essential and critical to success.
- **Personal responsibility** – individuals will take responsibility for, and control of, their own provisions for education, training and development, as well as for social insurance, pensions and healthcare.
- **Earnings** – helped by improved health, people will spread their earning capability over longer periods, with the concept of a fixed retirement age becoming irrelevant, as people determine their own working patterns from month to month and year to year, some working longer than others, some working more intensively and retiring earlier, some in education for longer periods and at various stages throughout their lives.

This picture of the future world of work is considered highly probable because the pressures from which it will develop are in existence and their effects are evident today (see Table 8.1).

In December 1998, a deal was struck between the workers and management at BMW's car plants in Europe, which is likely to pave the way for many other industries. At the heart of the new arrangement is a flexible working pattern that promises to deliver improvements in operating performance, and responsiveness to variations in market demand and in the utilization of plant and facilities. The key elements of BMW's flexible working practices (Fig. 8.1) comprise the following.

Table 8.1 A shift in the way work will be perceived

Old way	New way
Work is full-time employment, doing one job for one employer	Work is various contracts of varying durations each for a defined output with possibly several in operation at one time
Working hours are fixed, usually 40 hours in a working week	Working hours vary by day and by week to suit the achievement of a defined output
Success at work is built upon age, gender and length of service	Success at work stems from innovation, speed of response and flexibility
Investment in work is local or national	Investment in work is globally mobile
Big businesses are beautiful	Small businesses are faster to respond

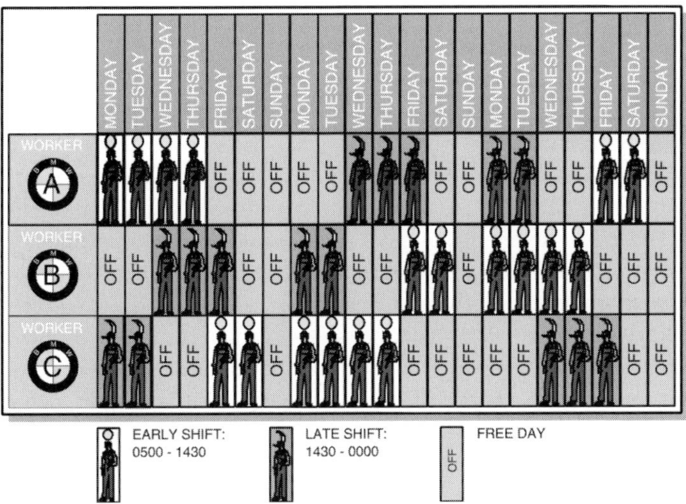

Fig. 8.1 The BMW flexible working pattern.

- Staff will work 35 hours per week, 2 hours less than they previously worked, which will be spread over four days rather than five days.
- The length of the 'production day' will be extended to 19 hours, based upon a three-week shift rota system, which will enable factories to operate at over 100 hours a week.
- When production is at its peak, staff will be able to work longer hours and when production is slack, staff will be able to take time off.
- Instead of overtime payments, staff will each have a 'Time-off Account' in which extra hours worked will be credited, to be repaid with days off when production is quieter.

We consider that it may not be long before similar flexible working principles are adopted by other businesses, both manufacturing and service based.

Evidence of the impact of this 'new order' can be seen in the way jobs have changed (Table 8.2). Work conventions of the comparatively recent past are being questioned if not overturned. The work patterns of individuals are changing from a fixed 'job for life' to the concept of 'life-long learning' where several changes of career in an individual's work-life will not only be commonplace but expected.

Consequently, in the future, work will be much more about delivering to a specification

Table 8.2 The changing nature of jobs (source: *Redefining Work – An RSA Initiative*).

Old jobs	New jobs
Banknote printer	Smart card designer
Bank clerk	Call centre agent
Company worker	Portfolio worker
Training manager	HR development consultant
Steel worker	Control technician
Draughtsman	CAD operator
Ambulance driver	Paramedic

than about working a pre-determined quantity of hours in a specific location. In turn, this will require managers to learn new skills to address the needs of the virtual organization. The need to change management practice will apply equally to facilities managers as it will to business managers. Business managers will need to develop skills in managing a network of people who do not work exclusively for one business or in one location. Facilities managers will need to embrace skills relating to all support services not just those appertaining to buildings, as has been the tradition. These skills will include ICT (information and communications technologies) work settings in 'remote' environments such as telecentres, the home and hotels; and the use of technologies to enable cost effective management of these remote resources.

Many workplaces of the 1990s have been designed as responses to old approaches to work and, in so doing, fail to take account of the present needs of people and businesses, let alone their future needs. This will have to change. But the required change has to be predicated on a real understanding of the needs of people, work processes and business, in the knowledge that what are regarded today as innovative and challenging working practices and relationships, will be commonplace in 25 years time.

8.2 The changing face of the workforce

As work changes, so too will the workers, and vice versa. Progressively, the world of business will move towards highly fluid 'employment' arrangements in its search for the means to be, and remain, flexible in response to the needs of customers and the changing dynamics of the marketplace. This fluidity could take many forms, but what they all share is the dynamic component of self-reinvention. In this process the business will continuously reappraise the composition of the workforce, its shape, location and the means that are put at the disposal of its people to perform their roles effectively. By way of example, Bayliss (1998) quotes how the size and shape of the workforce has changed within the steel manufacturing industry. The steel that was produced in the 1970s by the hard labour of 4000 men, is produced in the 1990s by a single plant half a mile long staffed by 11 technicians. Consequently the skills required to produce the steel have changed from 'blue collar' manual labour exercising their craft in close proximity to the foundry's furnaces, to 'white collar' technical, knowledge-based skills applied in a remote monitoring and control centre. The industrial landscape of much of the UK is a testament to these changing ways of working.

Many businesses will be small in size, often operating as a core around which a varying group of independent specialists are contracted to provide their services. Progressively large scale businesses with multi-layered pyramidal organization structures are giving way to flatter team-based structures, where the focus is increasingly upon the worker taking responsibility for his or her work output rather than it resting with a 'manager' or 'supervisor'. The responsibility will not only be for their 'own' work but also for how their contribution effects the combined efforts of the 'team', resulting in a shared responsibility for the team's output. As businesses develop this approach, we will see responsibility being gradually expanded as the individual is expected to take a shared responsibility for the business's output. Clearly, in small businesses this has always been the situation. However, evidence is pointing to this approach of management of outputs being applied in large businesses, particularly in the financial services industry, where the price of success or failure is often shared between many, and where the ultimate price of failure is shared by all.

For most businesses this may seem new and radical, for a very few it will seem obvious and the best way. One such business is the FI Group, which was founded by Stephanie (Steve) Shirley in 1962, as a small business called Freelance Programmers, which provided a computer programming service, which she ran from her own home. By the early 1990s it had grown to 'employing' 1100 workers, with a turnover of over £20 million. As if that was not surprising enough, over 70% of those workers work from their own homes linked together by an electronic network, and over 90% of them are women, a striking contrast to many older businesses in more traditional industries.

It was Henry Ford who said 'It is not companies who pay wages, but customers'. So where responsibility for decision making is devolved to workers who interact directly with the customer, rather than, as previously, being the role of 'senior management', it should not come as any surprise that these workers will be seeking appropriate recompense. In addition to workers seeking rewards commensurate with the levels of responsibility being entrusted to them, they will also require, in fact we believe demand, control over their means of producing the expected output. One facet of this control will manifest itself in workers demanding control over their work environment as a means of enabling them to be as effective and hence productive as they can be.

In *Beyond Workplace 2000* (Boyett and Boyett 1996), Joseph Boyett describes his vision of the new American business – sometimes called the learning organization, the virtual corporation, or the agile enterprise – in terms of the six characteristics possessed by these innovative organizations.

1. Production and service delivery processes are designed such that they can be reconfigured at a moment's notice to produce new and more sophisticated products or services, or to respond to changes in levels of demand for products and services. The business is totally flexible, and process innovation – the creative reinvention of how work is performed – becomes as important, if not more important than, product innovation.

2. Products and services are designed such that they can be reconfigured to a specific customer's requirements at the time of sale or order. Customers are given an almost limitless range of choices, allowing them to create a unique product or service, should they so require. The customer becomes actively involved in the process of creating the product or service, and in so doing is no longer just fulfilling the role of a consumer of value, but is contributing to the creation of value.

3. Products are designed for upgradability so customers can just replace component parts, instead of buying a new version of the product and throwing away the old one. In effect, the old product is remanufactured by switching out modular components.

4. Increased emphasis is placed on attracting and retaining knowledgeable, empowered workers who can use their individual talents and their access to a wide range of information to increase significantly the business's flexibility, responsiveness and level of innovation. The new organization values its workers as assets, more so than its equipment, as the workers possess the knowledge (the collective learning of the organization) that makes the business unique, setting it apart from its competitors.

5. It is more apt to offer products and services rather than solely one or the other. Products essentially become a platform for providing an ever-evolving set of value-adding services to customers and users.
6. The business does not try to do everything itself. Instead, it establishes a network of relationships with other organizations, in some cases even competing businesses, so that it can readily and speedily access specialized know-how on best practice operations, wherever they may be located, in order to create high-quality, low-cost, new products and services for customers on demand.

As businesses continue to re-invent themselves in the pursuit of an answer to 'what is our core business?', this will lead to the widespread adoption of the virtual organization. For many businesses the transition will not be a quick one, but a progressive metamorphosis – evolution rather than revolution. However, this does not mean that the composition of the workforce will not change in the short-term. On the contrary, as is already widely evident, businesses are forming new groupings almost by the day. Strategic alliances, joint ventures, partnering, contracting and sub-contracting, licensing of work processes and franchising, all of these are borne out of businesses' desire to focus on 'what we do best', while putting the rest of the operations in the hands of others for whom that part of the process is their core business (Fig. 8.2). Facilities managers are familiar with this approach, since the 1980s and 1990s have seen enormous strides taken in the outsourcing of facilities management services, regrettably not always with a beneficial outcome.

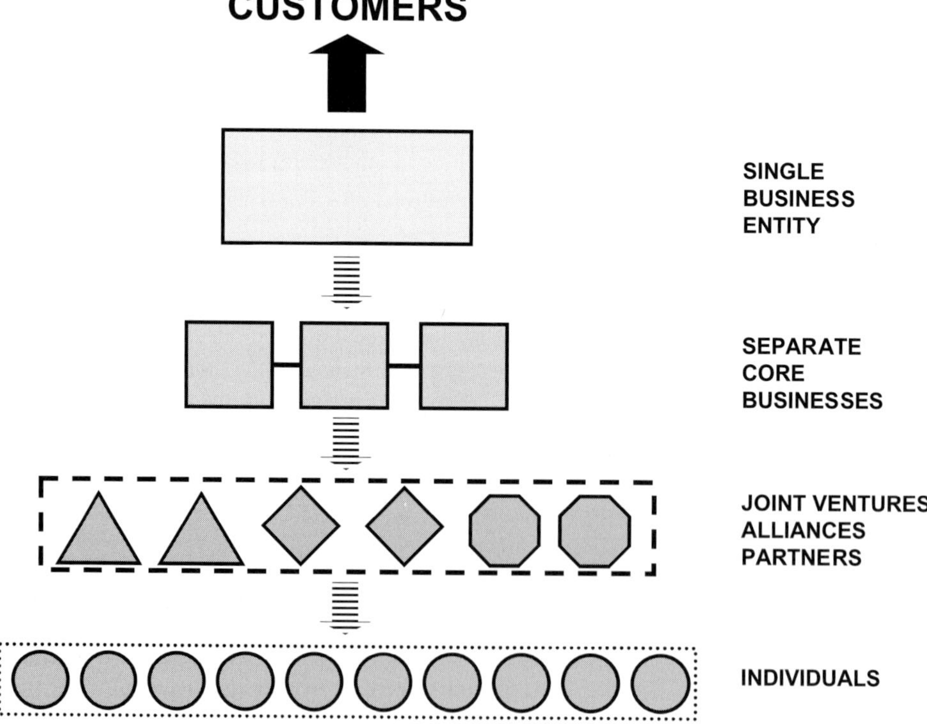

Fig. 8.2 The development of business shape.

Progressively, these multitudinous alliances will evolve more and more towards the virtual organization, where the people possessing the knowledge and skills necessary to create the products and services required by the customer, operate as a network of groupings of various types and sizes, in many locations. This dynamic, however, is unlikely to stop there. Rather than businesses solely moving from one state to another, the metamorphosis will itself be varied as businesses retain elements of their 'former selves' where to do so is consistent with good business practice. In this arrangement, core employees will work alongside a wide variety of workers who have widely different 'contracts of employment' with the business, some formal, some informal. This is likely to include freelance workers (both full-time and part-time), secondees from customers and suppliers, as well as workers from 'partner' businesses (Fig. 8.3). Therefore, when forming work environments it is of crucial importance for the effective working of the 'business team', that nothing is done that exaggerates the differences between team members.

With so many constituent parts, the business (which is what we shall continue to call the enterprise ultimately responsible for meeting the needs of the customer) will have to manage the contributions of the many and varied groups of workers from a diverse range of locations – the virtual organization. Moves towards the virtual organization will provide opportunities for previously disadvantaged sections of the workforce to be able to compete for work on a more equal basis than they have in the past. Not only will this include groups such as the physically disabled and the socially disadvantaged, who may be precluded from travelling to some work locations, but it will also enable all workers to be considered and participate on merit.

8.3 Lifestyles

Changing patterns of lifestyles are affecting us all. From environmentalism and communications, to health and education, balancing work life with home life will be an increasingly challenging prospect. While variable working patterns will be seen by some as a benefit, many people are likely, at least initially, to see this as a threat in the form of

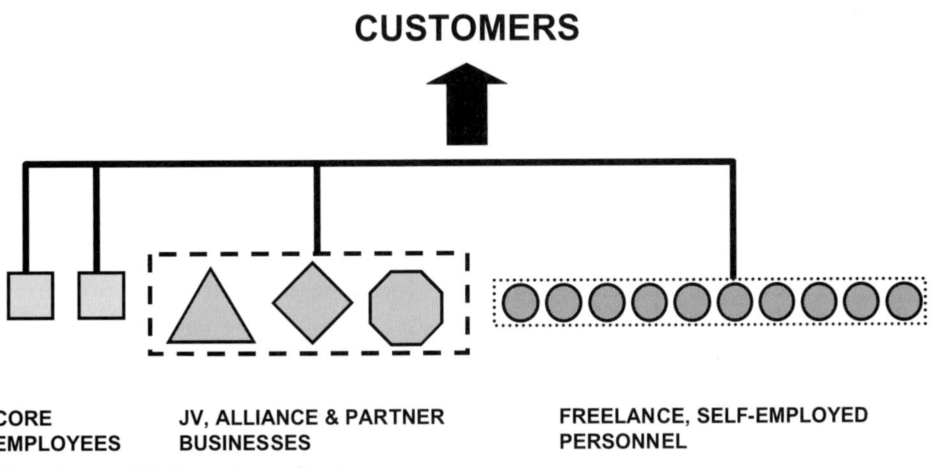

Fig. 8.3 A possible future shape of business.

insecurity of income rather than as an opportunity to balance the different facets of their life.

The concept of full employment, so often the battle cry of political candidates seeking election, will no longer be an obligation provided by, or expected from, employers nor is it ever likely to be in the gift of governments. Full employment will be redefined as an individual's ability to fill the amount of time they choose to allocate to work, with employment from one or more sources. People will take on personal responsibility for their training and development rather than expecting their employer to be the custodian of their career and skills enhancement. Already from an early age children are being introduced to what was, until comparatively recently, considered to be the 'adult world' of computers. The move has, in some instances, been so great, that it can only be a short while before the traditional three 'Rs' of reading, writing and numeracy, are increased to four by the addition of 'VR' in the form of computer literacy. Keyboard skills, so vital for success in the 'computing age' will soon be rendered obsolete by voice recognition systems. However, until that 'age' arrives, keyboard skills and general computer literacy will be many workers' route to 'full employment'.

If people can access almost limitless amounts of information on the Internet, and even get married over the Internet, then there is surely little impediment to apparently simpler tasks in life, such as shopping. In what promises to be the most significant revolution in high street clothing sales since the introduction of the measuring tape, retailers are planning to introduce virtual reality (VR) to assist the hard pressed shopper. Body scanners create a three-dimensional computer image of the shopper, which will enable them to try on 'virtual clothes' to achieve the desired look and perfect fit. Developed initially to aid medical and surgical treatment of patients, this application – being supported by major retailers such as Sears, C&A, Next, and Bhs – is clearly intended to improve current retail practice. It is, however, easy to see how the use of such technology could expand the already burgeoning home shopping services available via television and the Internet, with a potentially dramatic impact upon the face of the high street.

The pressures for improved environmental responsibility can only increase, as society comes to terms with the consequences of its actions. Reduced travel in general, and reduced business travel in particular, will be 'encouraged' by governments around the globe, in their drives for energy conservation and reduced levels of pollution. Reduced travel through the use of technologies such as e-mail and video conferencing, may hold the answer to the problems associated with the concentration of business and government headquarters in capital cities across the world. The extent to which the rapid growth in electronic and satellite traffic will change the economics of location is uncertain, but the motivation is likely to be quite different from its current drivers. We can already see some evidence of decentralization of 'backroom' functions to other locations by some major business sectors like banking and financial services. Such moves are not only beneficial in terms of savings in the facilities operating costs to the company, but also have the effect of improving the lifestyle of their employees through less time spent travelling to and from work, and perhaps, better equipped workplaces. In the field of education, the new technologies in IT have already spawned a rapid growth of distance-learning packages by universities and colleges, where remote students only 'meet' during specially convened e-mail 'chat-lines' or via video conferences.

Too frequently, corporate stress is addressed by businesses (consciously or unconsciously) passing on the problems to individual workers. For example, in one organization we encountered, the hurried introduction of new flexible styles of working had more to do with the business's failure to forecast its accommodation needs than it was to improve productivity or the quality of life for its workers. Here, the use of non-dedicated workplaces was perceived by staff as the removal of a benefit for the individual, where all the gains were in favour of the business. When working with clients on the introduction of home-based working, we have frequently encountered concerns expressed by employees who see the move to home working as the business passing its workplace problems on to the individual in terms of the provision of workspace, utility services and even catering.

There is much evidence to support the view that people who work from home are working longer hours and for longer unbroken periods. This, when coupled with the uncertainty of future employment (often itself caused by the introduction of home working), can give rise to higher than average stress levels being experienced by the home worker. Yet the personal attraction of working from home for many people is seen as being the lowering of stress, as they are more able to reconcile their work and home lives (Fig. 8.4). The key component appears to be choice. If an individual is given the ability to choose where and when they work, then they are likely to experience much reduced levels of stress.

A UK Department of Employment survey in 1987 discovered that the rate of pay for one fifth of all homeworkers covered by their survey put them in the top 10% of all earnings. So, as Charles Handy puts it, 'Homeworkers are out, telecommuters are in' (Handy, 1995, p. 86). Telecommuters have choice; they can chose to work before commuter trains start running or long after businesses' workspace is closed. They commute by telephone and e-mail, can move house without moving job, can take time off to enjoy the unpredictable brief spell of good weather or to celebrate some special event, without having to obtain anyone's permission.

Fig. 8.4 Working from home, for some or all of the time, will be a feature of many people's lifestyle (source: BT Corporate Picture Library; a BT photograph).

Such factors and the apparent lack of structure, have led some people to resort to what appear to be 'odd practices', in their attempts to create a clear distinction between work-life and home-life. There was the man who, at 8.30 am each morning, went out through his front door, walked clockwise all the way round his house (come rain or shine), re-entered the house through the front door and went to work in his home/office. The action was reversed at the end of the working day when he set out to walk anti-clockwise around his house before returning home 'from' work.

Flexible patterns of work can be of great benefit where 'packets' of work activity can be arranged to suit time-scales of family life. Finishing work at 4.30 pm rather than 5.30 pm or more commonly 6.30 pm, enables some workers to spend 2 to 3 hours with their young families, before sitting down to 1 or 2 hours distraction-free work at home after the children have gone to bed.

As businesses continuously strive for ever lower costs of production, salaries and other conventional 'pay packet' benefits will be less of a distinguishing feature between one business and another. In such situations the work environment, wherever it may be located, has the potential to become a business differentiator aiding recruitment and labour retention, through demonstrable affinity with the needs of the individual worker and in support of his or her quest for personal productivity.

8.4 The impact of technology and systems

The single most frequent failure in the history of forecasting has been grossly underestimating the impact of technologies. Because it is all around us and apparently continuous, the pace of IT development is hard to comprehend. In attempting to place it in some context, Charles Handy (1995, p. 14), makes the comparison, 'If the automobile industry had developed as rapidly as the processing capacity of the computer, we would now be able to buy a 400 mile-per-gallon (142 km/litre) Rolls Royce for £1'.

Advances in information and communication technologies through the 1980s and 1990s have been a driving force of change in many economies across the globe. IT development and its by-products have impacted on all aspects of our lives, and will continue to do so in the future. Short of an apocalypse, accelerating technological innovation will continue to shape the fortunes or constrain the development of businesses and nations alike.

Professor Peter Cochrane, Head of Research at BT Networks and Systems, describes (Bayliss, 1998) his view of what technology means for work and, by implication, what it means for the workplace.

1 Work is no longer a place – it is an activity that can be conducted anywhere.
2 Teleworking is not something to talk about it is something to do; a necessary and vital component of modern business life.
3 The world is already dominated by bits; information now makes more money than reshaping things made of atoms.
4 In the world of atoms, manufacturers no longer make cars and aircraft, they assemble them. They are systems integrators, using a myriad of small companies and individuals who supply the components. In the world of bits, this model is even more powerful – it is called the virtual organization.

5 We continue to educate and train people for a world that has gone, let alone for the world that is dying.

6 Electronic working means no job for life, but a new job every few weeks; multiple employers and continuous education and training are already the norm; things will get faster as technology removes distance and raises productivity.

7 Government, the law and social systems take months and years to react to change; modern companies work in hours and days.

8 In the past, managers controlled; in the future, there will be fewer of them, and they will be facilitators, communicators and doers, managing multiple operations on the move.

9 The biggest threat to any company is the CEO and Board who do not use, deny or do not understand the implications of IT.

10 Electronic working does not mean putting existing processes on screens; it means changing the nature of your business.

Sir Iain Vallance, Chairman of British Telecom, aptly sets the context for the central role of technology in the workplace of the future when he says, 'the competitive edge will be gained by those corporations giving their people the most natural access to new technology to blend workstyle and lifestyle.' Duffy (1995, p. 22), predicts that in the future 'a new type of office will be in regular use, creating an environment in which people are able to work intuitively, and organize information and interactions on a human scale without the barriers of hardware, software, time or space'. During the 1990s, a number of technological innovations were emerging that are likely to alter fundamentally the practice of work and the role of the work environment. These included the following:

- **Miniaturization of products** – IT products will continue to get smaller and have lighter weight while having enhanced quality of image and visual clarity, contributing further to improved mobility and flexibility.
- **Document image processing (DIP)** – is rapidly being enhanced, enabling those whose work processes continue to be paper-dependent to work remotely and simultaneously without compromising the integrity of the original document or their work activities.
- **Collaborative working applications** – further developments in groupware applications are evident, to assist groups of people, wherever they are located, to operate in a collaborative style of team working, through the use of technologies such as Lotus Notes, video-conferencing, data-conferencing and the Internet. When combined as an integrated system, these technologies – comprising messaging database, work-flow management, meetings package and diary scheduler – will enable meetings to be planned, convened, held and recorded, in a wholly virtual environment, where the users will experience improved performance through immediacy of contact, free-movement of information and hence increased personal as well as corporate performance.
- **Wireless networks** – have the capability of freeing-up the workplace and making reconfigurations speedier and less costly than in the past. Although currently operating at much lower speeds than hard-wired networks, as improvements in performance are developed, then the wireless LAN will be a common feature in almost every workplace.

- **Voice activated systems** – reducing, if not eliminating the need for typing, will enable a more immediate interface with technology such as e-mail that can be 'read' to its recipient. These systems will create different environmental needs in workspace, possibly requiring the use of headsets through which telephone communications will also be conveyed, or 'head-up displays' incorporating video transmissions and VR images; or possibly the use of semi-enclosed booths similar to telephone kiosks.
- **Light emitting polymers** – discovered in Cambridge, UK in the 1980s, light emitting polymers (LEP) are very thin and inexpensively produced 'plastics' that can be persuaded to pass an electric current which, when it comes in contact with impurities added to the polymer during manufacture, causes the polymer to emit light. LEPs open up the prospect of large-scale, low-cost projection of images, wrap-around visual displays and 'video wallpapers' that could add a significant new dimension to meeting rooms and collaborative working environments, as well as the potential for purely decorative displays.

According to British Telecom (BT), we will soon see even greater developments in telecommunications and related technologies than we have seen in the past. In their forecast of the future *BT World Communications Report*, BT advises that the race for the third generation of mobile telephones is now well underway. So much so that in many countries mobile telephone subscriptions have overtaken subscriptions for new fixed lines, which could soon make the fixed line a thing of the past. As BT puts it, 'Voice, fax, data and moving pictures are transported over the same telecom links. The next step will be to merge the computer and the phone, to free an individual from a fixed location so that voice and data calls can follow you around.' So, with the expansion of the use of the Internet, migration to the third generation of mobile communications, UMTS (Universal Mobile Telecommunications System), becomes an almost inevitable *must*, in terms of data traffic, graphics and video images.

Telecommunication and IT advances are allowing organizations to compete by 'virtualizing' the nature of their businesses. Many businesses will develop in whole or in part as virtual organizations, with their workforce linked to them by means of electronic 'umbilical cords'. This is happening because of three developments.

- Work is becoming abstract and information intensive.
- Teams are created and dismantled frequently, in response to varying business needs.
- The place of business is in electronic media, not a physical location.

The proliferation of effective and usable electronic media has allowed a large percentage of work activities to become location independent. For example, soon most of people's dealings with central and local government will be largely via electronic systems. Such initiatives will have profound implications for the locus of work, and resounding consequences for urban property markets.

Increasingly, workers will be able to be globally mobile without leaving home. In the software industry, major international companies in North America and Europe are already employing inexpensive programming expertise in India and Eastern Europe, where the difference in labour cost can be a factor of ten. The unstinting drive to reduce the cost of production will see continuing and progressive expansion of technology into spheres long thought of as requiring, if not depending upon, human intervention and action. The growth of big call centres, so much a feature of the 1990s, will be all but gone

Fig. 8.5 Given the comparatively short life expectancy of call centres, it is prudent to design them so they can be adapted for other uses such as at Orange plc, Darlington.

as new channels to market via home-based interactive technology are developed to replace them. Call centre agents working from home, or from places close to home rather than in aggregated central work buildings, will become the norm. Paid for by short-term gains businesses achieved during their brief period of competitive advantage, what legacy will call centres then leave to the stock of business buildings (Fig. 8.5)?

A continuously improving infrastructure is one of the imperatives for sustaining increases in business productivity and profitability. Consequently, issues relating to the possible application of technology to almost all aspects of business operations, and hence the place of work, will feature prominently in the strategic plans of every successful organization.

8.5 The workplace – a new geography

In one of his ever thought-provoking essays, Charles Handy (1996, p. 30), describes an incident he encountered when he was looking for a parking space and saw a man sitting in his car behind the wheel. 'Will you be long?' enquired Handy. 'About three hours, I think,' was the man's reply. Then, Handy saw the car phone, the portable computer and the fax, all on the seat beside the man, 'Good heavens', he exclaimed, 'you've got a whole office in there.' 'Sure', was the response, 'it's much cheaper parking here than hiring space up there', as he jerked his thumb at the high-rise office building behind them. The workplace has taken on a new geography (Fig. 8.6).

The workplace will be location-independent and user-driven, fully functional and supportive, and occupied on an intermittent pattern and for varying durations. As mentioned previously, the use of wide area networks (WANs), the Internet and video conferencing technologies make an expansive geography available, enabling people to work from any one of a variety of locations that could encompass premises of customers

Fig. 8.6 Some flexible working practices will reduce if not obviate the need for workspace (source: BT Corporate Picture Library; a BT photograph).

and suppliers, satellite buildings, transit locations and, of course, home. The principal objective being, to match the requirements of the tasks to be performed to the most effective time–space relationship that can be achieved. The chosen setting may be a desk within business premises; but equally it may be on a train or aeroplane, at home or in a hotel, in a café or a conference room, in fact anywhere that is conducive to the type and timing of work to be undertaken. It is easy to see how such a scenario could be construed as the demise of workspace – shop, office or factory. We do not believe this to be the case. While the overall *quantity* of workspace is likely to reduce, new styles of working will necessitate an increase in the *quality* of work environments and their location.

Internally, the micro-geography of the work environment will be every bit as fluid as the external macro environment. Becker and Steele (1995) cite the example of SOL, a cleaning company based in Helsinki, Finland, where, by the application of flexible, non-dedicated workplaces the business was able to support an increase in staff numbers based at the premises. Staff increased from 75 to 110 people without SOL having to make any changes to its existing environment or to add any new space, thus reflecting considerable savings in space costs as well as avoiding the disruption of reconfiguring the office that is usually associated with accommodating additional people.

The geography of work has local, regional, national and international dimensions. Technology recognizes neither borders nor time zones. Businesses serving the needs of their European customers can be serviced seamlessly via call centres located in lower wage countries in the Indian subcontinent or Pacific Rim countries. For large multinational companies, their production processes will be characterized by supply-chains that are international, exploiting comparative advantage in labour, raw materials, skills and technology, in order to service their global market.

The era of the building boom of the mid-1980s saw many very large scale office developments being created – based upon the aggregation of more and more people together, mostly into city centre locations – the future need for which is increasingly becoming questionable. Shared facilities will lead the drive for aggregation of building space but individual use of space will see a downward pressure as organizations fragment into smaller 'semi-autonomous parts' and technology encourages the further continuation of the trend, already in evidence, of remote working (Fig. 8.7).

Environmental realism, means no longer accepting that people should cause vast amounts of fossil fuels to be consumed as they travel to their place of work, when the activities can be conducted in dispersed satellite locations near home, where access can be achieved with less stress, quicker and while expending less energy. The triple whammy goes quadruple when, in the home environment, the worker is more productive than he or she would have been in the central workplace.

As mentioned in earlier chapters, enlightened organizations recognize the value of providing their staff with facilities and services that meet the needs of the 'whole person' and not just the 'worker'. Incorporating sports and welfare facilities within the workspace, is a laudable objective, where their anticipated use can cost-justify such initiatives. In providing fully supportive buildings however, care needs to be taken to ensure that they remain responsive. The more 'specialized' they become and the more bespoke features they incorporate, such as sports facilities, the more limiting may be the scope for future adaptation and disposal. Once the alternatives, such as membership of local clubs, or sharing the facilities of a neighbouring business, have been exhausted, and the need remains well proven, then provision of dedicated facilities may be appropriate. Can the facilities be cost-justified on their own, or could they be justified in any other way?

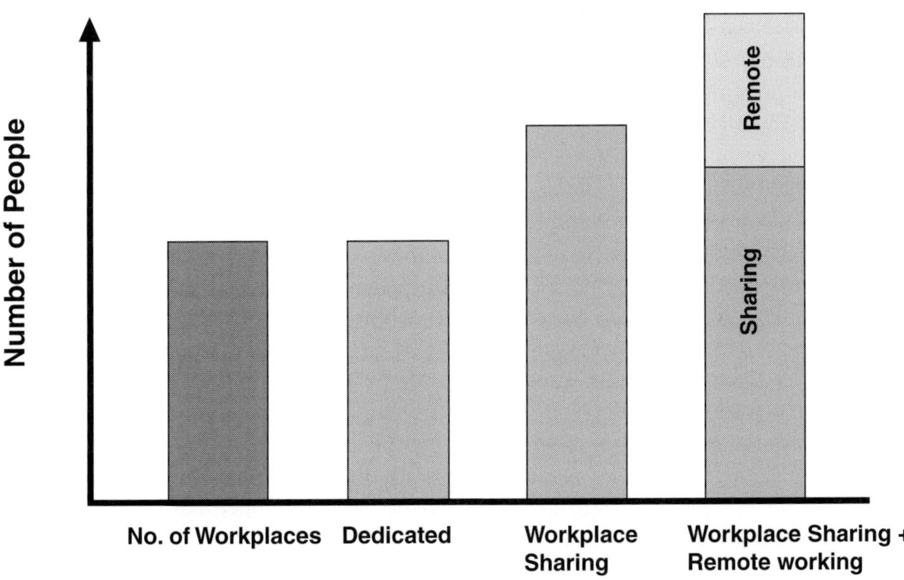

Fig. 8.7 Changing from allocating workplaces on a dedicated basis to more flexible approaches, can enable more people to be supported by the same number of workplaces.

Compare the difference in the approach taken at two British office buildings constructed over 20 years apart. At Willis Corroon's offices at Ipswich (designed by Sir Norman Foster in the early 1970s), a swimming pool was included as part of the staff recreational facilities, which eventually became redundant with changes in user desires, lifestyles and the availability of other local facilities. By contrast, at the administrative centre of The Scottish Office at Victoria Quay in Edinburgh (designed by RMJM Partnership in 1993) which also includes a swimming pool within the building, the surplus heat generated from the pool, instead of being vented out and wasted, is recycled through a heat exchanger to contribute towards the ongoing heating of the building. Here therefore, the facility is not just a recreational feature but is also an integral part of the building's environmentally conscious heating system.

Case Note 8a

Oil company, Midlands, England. In 1992 the new European head office of this oil company was designed on a bespoke basis for the company as a single occupier, which the company progressively occupied to full capacity. Within 3 years the business needs for workspace in that location reduced leaving parts of the building vacant. However, because the building was designed in discrete but interconnected modules, the oil company was able to mitigate its running costs by letting a portion of the building to a car manufacturer. Not only does the tenant's rent offset the oil company's property costs, but by sharing common support services, such as catering, the oil company is also able to defray some of its facilities operating costs, while the tenant business is able to avail itself of on-site services much more cost effectively than if it had to establish its own dedicated services.

Case Note 8b

Tasco Europe, Glasgow. Tasco is a new kind of organization, in a new area of business operating with new work styles. Formed as a joint venture between oil company Shell and accountants Ernst & Young, Tasco provides accounting and financial services, initially for Shell Group companies and eventually other multinational organizations across Europe. Starting with an initial 120 personnel, Tasco anticipates expanding rapidly to over 400. However there was a clear business objective that the expansion should not be achieved at the expense of compromising an effective work environment. On the contrary, the workspace is seen very much as an enabling facility in which initiative, responsibility and an open management style will permit the organization to adjust rapidly to meet varying market demands. In a climate of downsizing and outsourcing, the management of Tasco saw the opportunity, through the application of technology and new working methods, to provide professional services remote from their clients' premises.

To achieve the necessary responsiveness in the workplace, a structure to the environment was created where elements were classified as either 'hard fixes' or 'soft fixes'. Hard fixes, comprising those elements that would remain fixed for the duration of the organization's use of the premises, were kept to a minimum. Most elements were developed as soft fixes in order that they support the impromptu and varying needs of the business and its people. The resulting workspace is a bright, lively and fluid work environment in which the workers have control of workplace settings and the means of adaptation to meet their changing needs. Not for Tasco the traditional enclosed workplaces typical of so many professional services organizations, these were seen as constraining collaborative work practices – instead a rich mix of workplace settings is formed from movable hardware that supports different workstyles, such as:

- 'movable work walls' – floor to ceiling whiteboards that can be moved around;
- 'knowledge room' – quiet workspace where talking is not permitted; and
- spaces for 'stand up meetings' – equipped with high tables but no seating.

A key feature of the process employed to fashion the workspace was the 'up front' leadership demonstrated by the business's chief executive. She saw and grasped the opportunity to reinvent the work environment and the process by which it was created, similar to the way that Tasco itself was reinventing the process of business (Fig. 8.8).

8.6 Future trends – supply and demand

The early 21st century is an information age, where work is being redefined beyond traditional bounds of space and time. Communication and computing have created a global economy and business is responding with an increasingly virtual workplace that is having, and will continue to have, profound effects on all types of facilities. The value of the physical workspace is gradually being eroded while still being recognized as a powerful catalyst for promoting change in organizations. A major concern for facilities managers is how both new (and more often) existing building stock can be transformed, used, and managed over time. Becker and Steele (1995) in their book *Workplace by Design*, describe the need to create 'high performance workplaces' that are built on the basis that 'the workplace is a tool for achieving goals' and for 'putting space to work'. Predicting business futures is a tough assignment for even the sharpest of trend analysts. However, in the business of space, whilst there will be great uncertainties as to how the projected demand will be met by supply, there is a general consensus of demand indicators that are likely to shape the workplace of the future. These future workplaces are likely to be a result of responses to a number of diverse issues, including (Duffy, 1995) the following.

- The need to reduce costs and to use resources, particularly people, just when and where they are needed.
- Concentration on core business activities and not on what will be classed as diversionary, peripheral and non-core activities.
- The global marketplace in which product life cycles are shortening and new product development costs are increasing.
- The creation of virtual organizations which embody the whole supply chain, comprising the many diverse contributors required to fulfil a particular manufacturing process, and to deliver particular products or services.
- The shrinking pool of innovation and the niche players who think up new products and services, but who will not fit into corporate structures or locations.

(a)

(b) (c)

Fig. 8.8 (a), (b) and (c) A supportive work environment is essential to cope with the dynamics of new shapes of organizations, new forms of business practice and new styles of working – Tasco Europe, Glasgow (Hugh Anderson Associates).

- The availability of powerful information technology and communication services to control and manage effectively distributed and remote businesses, irrespective of geography or time.

On the supply side, property developers and owners will need to become more aware of the changing needs of businesses, and how new technologies will influence workstyles; the different ways in which businesses will use space in the future, and the consequential changes that will be required in the property market.

Buildings that are designed for single occupancy throughout their lives will be things of the past, or of short-sighted and short-lived property developers. In order to survive and flourish, the property industry will need to cater for the real needs of varying occupiers, their level and pattern of occupancy. Investors must recognize that the future performance of their property portfolios depends on their acceptance by users who will demand minimum operating costs *and* maximum flexibility.

In the office market, with the growth of serviced workspace populated by transient guests who visit to perform specific functions, offices are fast becoming hotels for work. The users of these buildings will expect a more comprehensive range of facilities and services than have traditionally been provided in such places. Like hotel guests, workers require facilities and services fully at their disposal while they are in 'residence', before moving on to another work location. All of which is likely to result in an overall reduction in the amount of workspace that will be required to meet the collective needs of businesses. This scenario has potentially devastating implications for property investment, pension funds and hence pensions. The consequences emanate from changes to the basis of planning offices, commercial and industrial buildings, where the amount of workspace provided has traditionally been a function of the number of people it has been necessary to group together under one roof. What does this mean for workspace? Potential factors affecting the nature of work-buildings and the way they are used, and hence their design criteria, include the following.

- **Educational buildings** – the widespread use of video- and data-conferencing enabling remote learning to be provided to all who need it by experts no matter where they are located, will question the need for centralized portfolios.

- **Office buildings** – video-conferencing links permit remote access to expertise (RATE) enabling businesses to provide specialist skills from dispersed locations rather than from concentrated 'centres of excellence'.

- **Factories** – in addition to increased automation in the factory workplace, fierce international competition will further squeeze product design to delivery cycles, resulting in demand for facilities that must go beyond the fundamentals of quality, reliability and cost; and even beyond flexibility, to what Stanley (1995) describes as 'agility' – the ability to make anything, in any volume, anywhere, at any time, by anybody.

- **Research facilities** – modern laboratories will require special considerations in handling sophisticated technologies and processes (electronic, bio-technical, and both together, as traditional discrete sciences become, at one and the same time, increasingly more specialized and more interdependent and integrated), demanding high standards of safety and security. To keep pace, research facilities must be designed and built to adapt quickly, inexpensively and with minimal disruption to ongoing processes.

- **Warehouses** – fully automated storage and retrieval of goods will enable products to be stacked to greater heights and hence higher densities, thereby minimizing the floor area required, enabling the building volume to be optimized, with almost no need for provision of human life support services.

- **Hospitals** – care in the community, particularly in remote locations, will be facilitated by technologies that allow remote diagnosis, monitoring and some treatment via computer and communications technologies, and a further application of RATE, as described above.
- **Libraries** – increasing amounts of published work will be available in electronic form (including via the Internet) as well as in hardcopy, allowing readers to select, read and return 'books' from varied remote locations.
- **Supermarkets and other retail outlets** – many goods will be viewed, selected and purchased remotely without the need for customers to enter the retailers' premises; the retailer will be able to adopt optimal storage techniques as referred to above, while minimizing the extent of its 'physical' shop window displays.
- **Homes** – will incorporate workspace, serviced with appropriate technology infrastructure, as an integral part of their design.

Buildings typically take 2–3 years to develop, they are leased for periods of up to 25 years, sometimes longer, in which time businesses will have disappeared or radically changed shape. In the same time-scale, new technologies will be developed, new products manufactured, and new skills learnt. Some buildings can be adapted to cope, many cannot, or at least not without compromising the effectiveness of the occupying business.

In response to such dynamics, the Learning Building Group (LBG) study (McGregor, 1994) proposed that, in addition to workspace being flexible, its procurement should also possess less rigidity than has for long been the hallmark of the property industry in the UK and most of Europe. The principal issues the industry needs to address include the following.

- **Funding and investment** – as businesses are metamorphozing over shorter time-scales, then lease periods of 25 years and longer, are no longer relevant, let alone acceptable; consequently other varied, and variable, types of funding and investment arrangements based upon 'partnering' between developer–owner–occupier, need to be provided.
- **Highly adaptable environments** – possessing the attributes of adaptability, capability, compatibility, controllability and sustainability, as described in Chapter 6.
- **Whole environment score** – which rates the whole-life of buildings on the totality of their impact upon the environment, covering such aspects as the energy and water they consume (in development and in operation), the levels of contaminants they emit (in development and in operation), their adaptability for other uses, the extent to which they make use of recycled products and materials, and the extent to which the materials used can themselves be recycled.
- **Total workspace provision** – where the building, and all of its infrastructure, furnishings and fittings, is available for 'hire' via a composite lease or other form of commercial arrangement.
- **Total facilities management (TFM)** – the management of the building, and all of its infrastructure, plant, machinery, utilities, services, furnishings and fittings, their servicing and maintenance, is provided by a specialist organization.

- **Lifetime guarantee** – where TFM is extended to cover the upgrading and/or replacement of all facilities throughout the occupiers' period of residence in the building, to meet their long-term variable needs.

It is encouraging to note that many aspects of the LBG's objectives for the future of workspace procurement are now beginning to emerge as fully developed commercial propositions. It has to be hoped that those not yet progressed will be embraced before long.

Previously seen as the domain of the small start-up enterprise which could not afford a 'place of its own', the use of the business centre will be at the heart of alternative accommodation strategies for many businesses, as they adopt a virtual approach to workspace, as suggested by Becker and Steele (1995). The business centre, a flexible, 'pay as you use' facility, where space of different types together with administration and support services are available on demand and on a bookable basis (by the hour, day, week, month), need not necessarily be something that is detached from the organization. Some organizations will make use of such centres in what might be called their 'conventional' mode, that is instead of establishing a permanent and dedicated work environment, they will rent space and buy services on an 'as needed' basis. Businesses such as IBM, Ford and British Telecom were quick to realize the economic potential of using this type of workspace. However, a less conventional approach, which we believe more organizations will adopt, will involve setting-up a business centre within their own premises, which they offer to their own staff and departments to use in support of their 'flexing' needs. This model can be further developed by organizations, through establishing in their own premises, a business centre primarily for use by third parties, but from which their own staff and departments can also rent space and buy services. The management and operation of such business centre models would most probably be best outsourced to a specialist company, such as the agreement between Lloyd's and Regus discussed in Chapter 6.

A model for the progressive use and development of a business centre could incorporate the following three stages.

Stage 1. Participation in the shared use of an external, commercial business centre as an occasional base for staff working remotely.

Stage 2. Setting up within the organization's principal premises, a business centre as an internal flexible service for use by all staff and departments, those 'based' at the premises as well as itinerant personnel, in which they can rent space and buy services.

Stage 3. In addition to serving the needs of its own organization, the internal business centre is made available for use by third parties on commercial terms. But need this approach be only the province of office space? There appears to be much scope for the property industry to employ this approach in servicing the needs of other sectors of business such as manufacturing, retailing and education.

Global investment strategies assisted in some cases by government subsidies, enable businesses to move production from country to country, as they search for low costs of production and available labour markets. However, this does not always result in the success story so widely predicted. What can come easily, can go easily. Hyundai were attracted to Halbeath, in Fife, Scotland to establish their European centre for silicon chip production, with significant inducements and the advanced construction of a manufacturing plant that was built to accommodate their needs. Before the plant, which promised employment for over 1000 people was even completed, Hyundai decided in the light of a world glut of silicon chips and the economic pressures they were experiencing at home in Korea, that it was no longer appropriate to have a facility at that location, leaving an empty and unused building in its wake.

The vision for the future of work buildings put forward by Peter Smith, who between 1989 and 1997 was Chairman of the RIBA Environment and Planning Committee, is one of environmental minimalism. Smith (1998) considers that by 2050 'it is possible to envisage buildings that make no demands on the environment apart from their embodied energy.' He speculates that these buildings may be net exporters, rather than consumers, of clean electricity, and may be up to 90% recyclable.

8.7 Implications for the future of workspace management

Whether or not you subscribe to all, or any, of the possible aspects of the future reviewed in this chapter, one issue is clear – work is going to change, almost certainly irrevocably, in terms of the how, when and where. Consequently, workspace, which is a product of business needs, must also change. So what then is the role of workspace and its management? The issues we have considered in this chapter will raise a great many socio-economic questions, the answers to which are not found in this book. What we attempt to do here, is to describe 15 key *drivers* for business change and the *implications* these issues raise for facilities managers as they manage their business's workspace.

1. **Business volatility** – As economies of scale give way to economies of scope, the convergence of businesses, constantly changing working practices, and a varying composition of the workforce, will be common in most enterprises. **Implication** – Almost all workspace will be shared rather than 'owned' by individuals, work groups, functional departments, and even by businesses and government departments – leading to an increase in demand for high-quality, flexible workspace, available on a variety of flexible commercial terms.

2. **Business location** – As risk is diversified away from the main downstream producer, rather than working from one or a few locations, businesses will typically operate from a great many locations, progressively evolving towards virtual and varying organizations. **Implication** – The types of locations and environments used by businesses will increasingly diversify – central hub facilities, dispersed satellite premises, telecentres, business centres, hotels, transport hubs such as airports and rail stations, transportation systems themselves such as aeroplanes and trains and, of course, home environments. These in turn, will call for a diversity and depth of skills, knowledge and resources not currently possessed by many facilities management teams.

3. **Application of technology** – Technology will penetrate deeper into work processes not previously envisaged, and the intensity of use of this technology will dramatically increase. **Implication** – Although the density and complexity of the technology will similarly intensify, benefits will be derived from substantial convergence in hardware and the inter-operability of an ever-expanding range of software applications. This,

together with a dramatic rise in the numbers of people who are not only computer literate but for whom the use of technology is a 'natural' way to do things, will mean that workspace that cannot support this style of working will be rejected by business.

4. **Environmentalism** – Improvements in environmental responsibility, driven by both legislative and voluntary initiatives, will continue to exert pressure for improved performance of workspace and buildings as eco-systems. **Implication** – All facets of the design and construction of work buildings will be selected after a thorough assessment of their environmental impact and performance targets set for their lifetime operation. Facilities managers will therefore need to have available to them, reliable data about the operating performance of the buildings in their charge, robust indices against which this performance data can be compared, and effective means of exercising control over all aspects of the building's operations.
5. **Availability of technology** – Technology systems previously thought of as being purely the domain of business will be available and accepted in the domestic environment, such as: interactive multimedia terminals replacing computers and televisions, mobile phones replacing fixed line telephones, video conferencing replacing audio conferencing. **Implication** – Houses need to be built to meet the requirements of the home worker, through the provision of a workplace, LAN, ISDN lines and other facilities – the home worker will be seen increasingly as a well-defined and separately identifiable segment of the housing market.
6. **Work patterns** – Worldwide working online will be common, hence the pattern of the working day, week, month and year will vary for each individual and from day to day and week to week. **Implication** – There will be less distinction between week days and weekends, as people work at times that are convenient to suit their lifestyles, with the direct consequence of increasing the intensity of use of the workspace.
7. **Workspace availability** – Erratic and often unpredictable patterns of working will be the norm rather than the exception, as businesses focus upon responding to the individual needs of customers. **Implication** – Premises and infrastructure will need to be available 24 hours a day, seven days a week, to suit global business across different time zones, as well as the personal lifestyle choices of people as they balance their needs for work and home. Such requirements can only be met through the provision of more robust, resilient and controllable building systems than have been available in the past, which will necessitate changes in the way investments are made in business workspace.
8. **Premises operating hours** – Many public holidays will be replaced by personal holidays (individual or group) to suit people's lifestyle needs and the demands of work and deadlines. That is not to say that the total amount of holiday time will reduce, on the contrary, the total amount of time each person takes as holiday will increase, but the timing will be much less uniform and predictable. **Implication** – Maintenance and servicing of workspace and the workplaces they contain will need to be highly responsive so as to take advantage of every small window of opportunity available to them, in order not to disrupt business operations while still maintaining business systems at optimum levels of performance.
9. **Dispersed working** – Simultaneously, the effects of environmental pressures, the pervasiveness of IT and a re-evaluation of the quality of life, will combine to

encourage many workers to explore new ways of working, as a means of liberating them from the personal effects of extensive commuting, stress, unproductive time, and soaring costs. **Implication** – The numbers of people working *from* home as well those working *at* home, will increase and where neither of these is practical, extensive use will be made of other work-nodes such as satellite premises, telecentres, as well as the premises of customers, suppliers and partners.

10. **Composition of the workforce** – In attempting to bring together the best skills, experience and practices, businesses will move progressively from being dependent solely upon full-time employees, to engaging a range of different types of workers via many different forms of 'contract', most of whom will work predominantly in dispersed locations. **Implication** – The business will require a 'central hub', or possibly several dispersed 'hubs', where workers can meet, share experiences, collaborate and socialize. Consequently, workspace will need to incorporate more team and collaborative work settings, as well as more social spaces of different types and scales.
11. **Worker expectations** – Rising standards of education, together with higher levels of skill, empowerment and responsibility, will fuel the increasing expectations of workers to have work environments that truly support them in their endeavours as individuals and teams. **Implication** – Higher standards of functionality will be provided as the norm in work environments where the workplace and its support services are operating under the direct control of workers and not of facilities managers. In such a situation, the role of the facilities manager will be that of the 'enabler', marshalling the services required, keeping them under constant surveillance and continually seeking improved performance through the application of innovative techniques. This will, in turn, require the facilities manager to conduct frequent surveys of his or her customers (workspace users) to ascertain the level of their satisfaction with the work environment, its infrastructure and services.
12. **Workplace control** – Workers, individually and collectively, will be held much more accountable by the business for the quality and quantity of their output. This, together with a better understanding of the correlation between the performance of the work environment (in its most holistic sense) and the performance of its users, will mean workers will be much less tolerant than in the past, of defective and deficient workspace (Fig. 8.9). **Implication** – The provision of facilities services, the customization of workplace settings and the adaptation of workspace layouts, all need to be achieved much more readily than has been experienced previously. The facilities manager and his or her team, become custodians and providers of the means of production, the control of which is devolved to *their* customers. This is likely to require new forms of service delivery, provided via various and variable forms of contract, more expansive help-desk services embracing remote diagnostics, and the provision of the means by which individual control can be achieved, featuring as an integral part in the selection of work setting components. All of which will have profound implications for compliance with workplace health and safety and other increasingly stringent legislative obligations.
13. **Workplace performance** – Competitive business pressures, continuing an unrelenting drive for the lowest possible cost of production (for both goods and

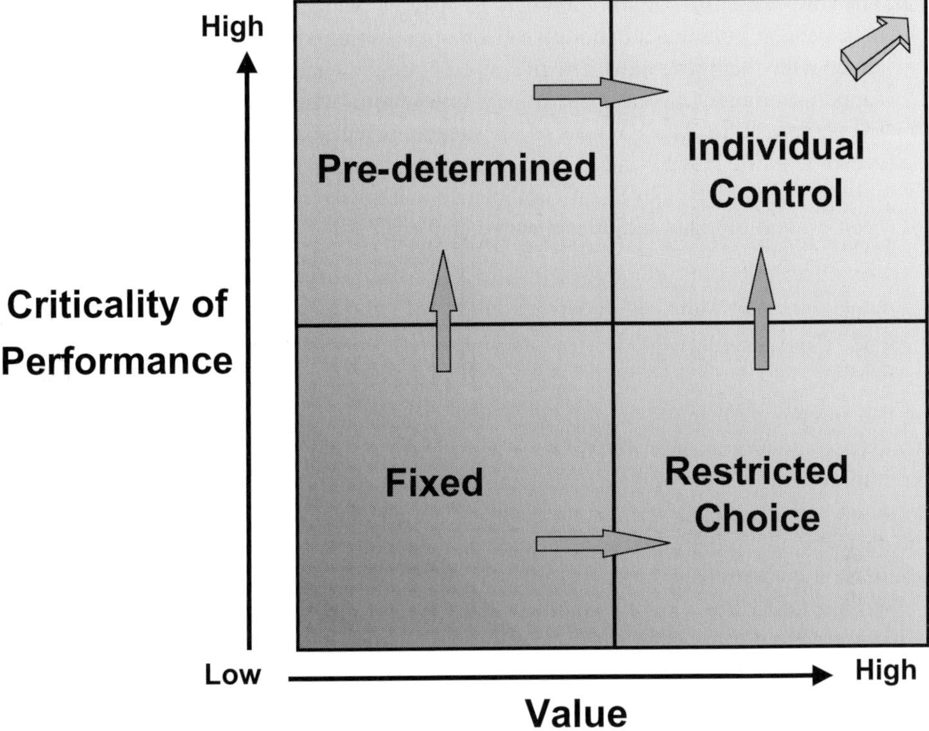

Fig. 8.9 As businesses make the linkages between the performance of the workplace and the performance of their people, devolving control of workplaces to individuals becomes inevitable.

services), will demand high levels of operating performance from the individual and the team, as well as the organization as a whole. **Implication** – The management of workspace will be as much about measurement, as it will be about implementation. The facilities manager will need to possess a deep understanding of the way the workplace affects the performance of people (individually and collectively), how the workplace is being used and why, and possess the ability to orchestrate changes in the way work is conducted, confident in his/her knowledge of the cause and effect linkages. By evaluating the performance of the workspace and linking the findings to recognized business metrics, the facilities manager will then be able to make a value added contribution to the business.

14. **Facilities management operations** – As the geography of the workplace changes, smaller cores of workspace will be required at the 'centre' of businesses, giving rise to many and diverse types of workspace being used in a wide range of building types in widespread locations. **Implication** – The critical mass required to sustain a dedicated and local facilities management team is unlikely to exist, so new initiatives will need to be developed in response to the changing scene. These will encompass methods and systems for remote operation, monitoring and diagnosis of facilities services; through collaboration, joint venture and sharing, the facilities management functions of neighbouring businesses will combine and rationalize in support of a range of local buildings and their occupants.

15. **The role of facilities management** – To be effective and remain so, workspace needs to be created from the collaborative efforts of all aspects of business, where IT and HR along with facilities management are combined to achieve an effective integrated support service for business operations. **Implication** – Facilities management expands to encompass the management of all support services as part of a business support group bundling HR, IT and finance along with the management of facilities. This will call for much higher levels of competence and business skill than traditionally have been evident in many facilities management practitioners.

8.8 Conclusions

In the foregoing chapters we have briefly charted the history and development of workspace and its management. We have described procedures and methodologies that can be employed in developing an understanding of the needs of business, its processes, premises and, most importantly, its people. The aim is to enable the facilities manager satisfactorily to address the supply and demand equation of the workspace needs of business, at both strategic and tactical levels.

We have also taken a brief look into what the future may hold for the management of workspace. From this glimpse it is clear that businesses' requirements for workspace will change in terms of quantity, quality, location, diversity and functionality, all of which will place greater and more varied demands on those responsible for its provision and its management. These demands will call for new skills to be learnt, new practices to be deployed and new procedures to be adopted by all of those servicing the needs of workplace users.

We have pointed to the need for attention to be given to the performance of the workplace, based upon the application of objective measures conducted within a context of customer orientated services. To maintain an effective role in the organization, the facilities manager will need to innovate new solutions to new problems based ever more on looking forward and much less on looking back.

In response, the property industry needs to develop a wide range of products for the market, which are likely to require new approaches to the way property is funded, designed, constructed, equipped and provided. The facilities services industry will need to shape new offerings based on more appropriate levels of service for businesses and their people, which are developed in recognition of the need to be value adding rather than resource consuming.

All of these requirements stem from the fundamental tenet – customer organizations are far less willing, let alone able, to tolerate work environments that compromise their business operations, than they have been in the past. Consequently, managers of facilities must make a major shift of mindset by recognizing that simply adopting coping strategies is insufficient as the focus of their actions shifts from resource and expenditure containment, to adding value to the business.

For this to be attained, facilities managers will need to innovate new ways of providing work environments and services, which will in turn require more flexible contract structures and service level agreements. These and other demands will necessitate facilities managers driving the property and facilities service industries, rather than vice versa. The focus must be upon effective performance of the workplace and its related services, and their direct correlation with the performance of their users and hence the overall performance of the business. In such an ambience, facilities management, with all of its diversity and multitudinous activities, must have at its nucleus – the *business of space*.

References

Bayliss, V. (1998) *Redefining Work* (London: RSA).

Becker, F. and Steele, F. (1995) *Workplace by Design – Mapping the High-performance Workscape* (San Francisco: Jossey-Bass).

Boyett, J.H. with Boyett, J.T. (1996) *Beyond Workplace 2000: Essential Strategies for the New American Corporation* (New York: Plume/Penguin).

Duffy, F. (1995) The future for offices. *Flexible Working*, **1** (3), 21–22.

Handy, C. (1995) *The Age of Unreason* (London: Arrow Business Books), pp. 14, 86.

Handy, C. (1996) *Beyond Certainty* (London: Arrow Business Books), p. 30.

McGregor, W.R. (1994) An integrated workplace process. *Facilities*, **12** (5), pp. 20–25.

Smith, P. (1998) Green buildings: the long-term payoff. In Edwards, B. (ed) *Green Buildings Pay* (London: Routledge), p. 132.

Stanley, J. L. (1995) Beyond flexible – tomorrow's agile manufacturing facility. *Facility Management Journal*, IFMA, May/June, 16–20.

Bibliography

Akhlaghi, F. (1993) The business of space. In: Paper presented at a conference on *Quality Concepts and Premises Management in Further Education*, Sheffield Hallam University, 30 April.

Allison, S. (1994) PowerGen's space-planning mission. *Facilities*, **12** (11), 11–15.

Apgar IV, M. (1993) Uncovering your hidden occupancy costs. *Harvard Business Review*, May–June, 124–136.

Apgar IV, M. (1995) Managing real estate to build value. *Harvard Business Review*, Nov–Dec, 163–164.

Arthur Andersen (1993) *Real Estate in the Corporation – The Bottom Line from Senior Management*. pp. 7–8.

Avis, M., Gibson, V. and Watts, J. (1989) *Managing Operational Property Assets* (Reading: Graduates to Industry).

Avis, M., Gibson, V. and Watts, J. (1995) *Real Estate Resource Management* (Wallingford, Oxon: GTI).

Balch, W.F. (1994) An integrated approach to property and facilities management. *Facilities*, **12** (1), 17–22.

Baldwin, R., Yates, A., Howard, N. and Rao, S. (1998) *BREEAM 98 for Offices* (Watford: BRE).

Barrett, P. (ed.) (1995) *Facilities Management – Towards Best Practice* (Oxford: Blackwell Science).

Bayliss, V. (1998) Redefining Work. (London: RSA).

Becker, F. (1981) *Workspace: Creating Environment in Organisations* (New York: Praeger Press).

Becker, F. (1990) *The Total Workplace: Facilities Management and the Elastic Organisation* (New York: Van Nostrand Reinhold).

Becker, F. and Steele, F. (1995) *Workplace by Design: Mapping the High-Performance Workscape* (San Francisco: Jossey-Bass).

Beitz, D. (1997) *Presentation at FMA Professional Development Program*, Brisbane, November.

Bon, R. (1989) *Building as an Economic Process – An Introduction to Building Economics* (Englewood Cliffs, New Jersey: Prentice Hall).

Boyett, J.H. with Boyett, J.T. (1996) *Beyond Workplace 2000: Essential Strategies for the New American Corporation* (New York: Plume/Penguin).

Brandt, R. (1993) What is strategic facilities planning? *FM Journal*, IFMA. May/June, 36–39.

British Institute of Facilities Management, *Best Practice Guide: Space Planning* (London: BIFM).

British Institute of Facilities Management, *Best Practice Guide: Business Continuity Planning* (London: BIFM).

British Institute of Facilities Management, *Facilities Management Measurement Protocol* (London: BIFM).

Byrne, J.A. (1992) Management's new gurus. *Business Week*, 31 August, 44–52.

Centre for Facilities Management (CFM) (1991) *UK FM Market Trends Survey*.

Champy, J. and Hammer, M. (1993) *Re-engineering the Corporation – A Manifesto for Business Revolution* (Nicholas Brealey) p. 17.

Chekijian, C. (1994) Shrink, shrink, shrink that space for big savings, increased profitability, shareholder. *Corporate Real Estate Decisions*, October, 7.

Chekijian, C. (1996) Demand management must focus on corporate objectives. *Corporate Real Estate Executive*, **11** (2), 36–37.

Crainer, S. (1996) That was the idea that was. *Management Today*, May, 38–42.

Creighton, J.L. and Adams, J.W R. (1998) *Cyber Meeting* (New York: AMACOM).

Crisp, V.H.C., Doggart, J.V. and Attenborough, M.P. (1993) *BREEAM 2/91 An Environmental Assessment for New Superstores and Supermarkets* (Watford: BRE).

Crosby, N., Gibson, V., Lizieri, C. and Ward, C. (1998) Market views on changing business practice, leasing and valuations. *Right Space: Right Price?* **6** (London: The Royal Institution of Chartered Surveyors).

Debenham Tewson Research. (1992) *The Role of Property – Managing Cost and Releasing Value* (Debenham Tewson).

Deloitte and Touche (1995) *Space Needs of Corporate Headquarters in America* (Chicago: Deloitte and Touche Real Estate Service).

DETR (Department of the Environment, Transport and the Regions) (1998) *Good Practice Guide 237: Natural Ventilation in Non-domestic Buildings* (Watford: BRESCU).

Duffy, F. (1983) *ORBIT Study* (London: DEGW).

Duffy, F. (1990) Measuring building performance. *Facilities*, **8** (5), 17–21.

Duffy, F. (1995) The future for offices. *Flexible Working*, **1** (3), 21–22.

Duffy, F. (1998) *The New Office* (London: Conran Octopus).

Duffy, F. and Tanis, J. (1993) A vision of the new workplace. *Site Selection* **38** (2) Industrial Development Section, 427–432.

Duffy, F., Laing, A. and Crisp, V. (1993) *The Responsible Workplace – the Redesign of Work and Offices* (London: Butterworth Architecture in association with Estate Gazette).

Edwards, B. (1998) *Green Buildings Pay* (London: Routledge).

Edwards, V. and Seabrooke, B., 'Proactive property management.' *Property Management*, **9** (4), 373–384.

Eley, J. and Marmot, A. (1995) *Understanding Offices* (Penguin Books).

Evans, M. (1994) The impact of real estate decisions on profitability. *Site Selection Europe*, March, 28–30

Evary, M. (1994) The impact of real estate decisions on profitability. *Site Selection Europe*, March, 28–30.

Flanagan, R., Norman, G., Meadows, J. and Robinson, G. (1989) *Life Cycle Costing – Theory and Practice* (London: BSP Professional Books).

Gale, J. and Case, F. (1989) A study of corporate real estate resource management. *Journal of Real Estate Research*, **4** (3), 23–34.

Gerald Eve (1997) *Overcrowded, Under-utilised or Just Right* (London: Gerald Eve).

Gibson, V. (1998) Changing business practice and its impact on occupational property portfolios. *Right Space: Right Price?* **4** (London: The Royal Institution of Chartered Surveyors).

Gibson, V. and Lizieri, C. (1998) The real estate implications of business reorganisation and working practice. *Right Space: Right Price?* **1** (London: The Royal Institution of Chartered Surveyors).

Gordon, A. (1974) Architects and resource conservation. *RIBA Journal*, January, 9–11.

Handy, C. (1995) *The Age of Unreason* (London: Arrow Business Books).

Handy, C. (1996) *Beyond Certainty* (London: Arrow Business Books).

Handy, C. (1997) Boring workplace, boring worker. *Management Today* November, p. 29.

Hiles, A. (1993) *Service Level Agreements – Managing Cost and Quality in Service Relationships* (London: Chapman & Hall).

HOK Consulting (1995) *Alternative Officing*. The HOK Facilities Consulting Report, p. 3.

Holmes, R. (1988) *A Systematic Approach to Property Condition Assessment in Whole-Life Property Asset Management*, Then, D.S.S. (ed.) (Edinburgh: Heriot-Watt University) pp. 140–150.

Hronec, S.M. (1993) *Vital Signs* (New York: AMACOM).

Jones, O. (1997) The service office – the future of flexible space? In: *Proceedings of the British Institute of Facilities Management – Facilities Management Into the Business Revolution*, Cambridge, 15–17 September, 31–38.

Joroff, M., Louargand, M. Lambert, S. and Becker, F. (1993a) *Executive Summary. Report of Phase One Corporate Real Estate 2000*. IDRF, pp. 7–8.

Joroff, M., Louargand, M., Lambert, S. and Becker, F. (1993b) *Strategic Management of the Fifth Resource: Corporate Real Estate. Report of Phase One: Corporate Real Estate 2000* (USA: The Industrial Development Research Foundation).

Kimmel, P.S. (1993) The changing role of the strategic facilities plan. *FM Journal*, IFMA, May/June, 24–29.

Laing, A. (1997) New Patterns of Work: the Design of the Office. In Worthington, J. (ed.), *Reinventing the Workplace* (London, Architectural Press), p. 36.

Laing, A., Duffy, F., Jaunzens, D. and Willis, S. (1998) *New Environments for Working: The Re-design of Offices and Environmental Systems for New Ways of Working* (London: E&FN Spon).

Leaman, A., Bordass, B. and Cassels, S. (1998a) *Flexibility and Adaptability in Buildings: the 'Killer' Variables* (London: http://www.usablebuildings.co.uk).

Leaman, A., Cassels, S. and Bordass, B. (1998) *The New Workplace: Friend or Foe?* (London: http://www.usablebuildings.co.uk).

Lindsay, C.R.T., Bartlett, P.B., Baggett, A., Attenborough, M.P. and Doggart, J.V. (1993) *BREEAM/New Industrial Units Version* **5/93** (Watford: BRE).

Marshall, H.E. and Ruegg, R.T. (1990) *Building Economics – Theory and Practice* (New York: Van Nostrand).

McGregor, W.R. (1994a) An integrated workplace process. *Facilities* **12** (5), 20–25.

McGregor, W.R. (1994b) Designing a 'learning building'. *Facilities* **12** (3), 9–13.

McLocklin, N., Maternaghan, M., Lowe, S. and Bevan, M. (1995) *Technology and the Workplace. Facilities Management*, Management Guide No 9 (London: The Eclipse Group).

Nourse, H.O. and Roulac, S.E. (1993) Linking real estate decisions to corporate strategy. *The Journal of Real Estate Research* **8** (4), 492.

Oakland, J.S. (1989) *Total Quality Management* (Oxford: Heinemann Professional Publishing).

O'Reilly, J.J.N. (1987) *Better Briefings Mean Better Buildings* (Watford: BRE).

Parshall, S. and Kelly, K. (1993) Creating strategic advantage. *Premises and Facilities*, August, 11–16.

Pélegrin-Genel, Élisabeth. (1996) *The Office* (Paris: Flammarion).

Peters, T.J. and Waterman, R.H. (1982) *In search of Excellence* (New York: Harper & Row).

Pratt, K. (1994) Introducing a service level culture. *Facilities*, **12** (2), 10.

Pratt, K. (1997) Current space use strategies and churn management. *Facilities*, **15** (9/10), 224–226.

Pringle Brandon (1996) *20/20 Vision: A Report on the Impact of Flat Panel Displays on Trading Room Design* (London: Pringle Brandon).

Propst, R. (1968) *The Office – A Facility Based on Change* (Michigan, USA: Herman Miller).

Quinn, J.B. and Hilmer, F.G. (1994) Strategic outsourcing. *Sloan Management Review*, Summer, 43.

Sager, I. (1995) Big Blue's white-elephant sale. *Business Week*, 20 February, p. 36.

Senge, P. (1990) *The Fifth Discipline: The Art and Practice of the Learning Organisation* (New York: Doubleday).

Smith, P. (1998) Green buildings: the long-term payoff. In Edwards, B. (ed) *Green Buildings Pay* (London: Routledge) p. 132.

Spedding, A. (ed.) (1987) *Building Maintenance, Economics and Management* (London: Spon).

Stanley, J.L. (1995) Beyond flexible – tomorrow's agile manufacturing facility. *Facility Management Journal*, IFMA, May/June, 16–20.

Thatcher, C. and Thatcher, G. (1996) *Reducing the Cost of Churn. Facilities Management*: Management Guide No 10 (London: The Eclipse Group).

The Henley Centre (1996) *The Milliken Report: Space Futures*. Commissioned by Milliken Carpets.

Then, D.S.S. (1994a) Asset management and maintenance – Part 1. *Facilities Management* **2** (2), 10–12.

Then, D.S.S. (1994b) Facilities management – the relationship between business and property. *Proceedings of EuroFM/IFMA Conference on Facility Management European Opportunities*, Brussels, p. 259.

Then, D.S.S. (1994c) People, property and technology – managing the interface. *Facilities Management*, **2** (1), 6–8.

Then, D.S.S. (1994d) Property as an enabling resource to business. *Facilities Management*, August, pp. 4–7.

Then, D.S.S. (1995) Asset management and maintenance – Part 2. *Facilities Management*, **2** (3), 7–9.

Then, D.S.S. and Fari, A. (1992) A framework for defining facilities education. *Facilities Management – Research Direction*. Barrett, P. (ed.) pp. 15–22.

Then, D.S.S. (1996) A study of organisational response to the management of operational property assets and facilities support services as a business resource – real estate asset management. Unpublished Thesis. Heriot-Watt University, Edinburgh, Scotland. UK.

Varcoe, B. (1991a) Proactive premises management – the premises policy. *Property Management*, **9** (3), 224–230.

Varcoe, B. (1991b) Proactive premises management – asset management. *Property Management*, **10** (1), 224–230.

Varcoe, B. (1996) A key to unlock the future. *Facilities Management*, **3** (5), 6–8.

Veale, P.R. (1989) Managing corporate real estate assets – current executive attitudes and prospects for an emergent management discipline. *Journal of Real Estate Research*, **4** (3), 16.

Veldon, E. and Piepers, B. (1995) *The Demise of the Office* (Rotterdam: Uitgeverij).

Vischer, C. (1996) *Workspace Strategies: Environment as a Tool for Work* (New York: Chapman & Hall).

Worthington, J. (1997) *Reinventing the Workplace* (Oxford: Architectural Press).

Zeckhauser, S. and Silverman, R. (1981) *Corporate Real Estate Asset Management in the United States* (Cambridge, Massachusetts: Harvard Real Estate).

Further reading and reference

British Council for Offices (1994) *Specification for Urban Offices* (London: British Council for Offices).

British Institute of Facilities Management, *Best Practice Guide: Business Continuity Planning* (London: BIFM).

British Institute of Facilities Management, *Best Practice Guide: Lighting for the Workplace* (London: BIFM).

Butler Cox (1989) *Information Technology and Buildings* (London: Butler Cox).

Crosby, N. and Lizieri, C. (1998) *Changing Lease Structures – an Analysis of IPD Data. Right Space: Right Price?* **5** (London: The Royal Institution of Chartered Surveyors).

Crosby, N. and Murdoch, S. (1998) *Changing lease structures in commercial property markets. Right Space: Right Price?* **2** (London: The Royal Institution of Chartered Surveyors).

Department of Trade and Industry (1997) *ISI Programme for Business: How Videoconferencing Can Work for You* (London: HMSO).

Handy, C. (1989) *The Empty Raincoat* (London: Hutchinson).

Health and Safety Commission (1992) *Workplace (Health, Safety and Welfare) Regulations 1992: Approved Code of Practice and Guidance L24* (London: HMSO).

Henderson, J. (1998) *Workplaces and Workspaces – Office Designs that Work* (Gloucester, Massachusetts: Rockport Publishers).

Peters, T. (1987) *Thriving on Chaos: Handbook for Management Revolution* (New York: Alfred A Knopf).

Peters, T. (1992) *Liberation Management: Necessary Disorganisation for the 90s* (London: Macmillan).

Pheasant, S. (1998) *Bodyspace – Anthropometry, Ergonomics and the Design of Work* 2nd edn (London: Taylor & Francis).

Piore, M. and Sabel, C. (1982) *The Second Industrial Divide: Possibilities for Prosperity* (New York: Basic Books).

Raymond, S. and Cunliffe, R. (1996) *Tomorrow's Office: Creating Effective and Humane Interiors* (London: E&FN Spon).

Ward, C., Crosby, N. and Lizieri, C. (1998) Using asset pricing models to value property interests. *Right Space: Right Price?* **3** (London: The Royal Institution of Chartered Surveyors).

Welsh, J. (1997) Green piece. *RIBA Profile*. April.

Williams, B. (1988) *Premises Audits* (London: Bulstrode Press).

Zelinsky, M. (1998) *New Workplaces for New Workstyles* (New York: McGraw Hill).

Index